D1272924

THE IP MULTIMEDIA SUBSYSTEM (IMS)

Session Control and Other Network Operations

Travis Russell

New York   Chicago   San Francisco
Lisbon   London   Madrid   Mexico City
Milan   New Delhi   San Juan   Seoul
Singapore   Sydney   Toronto

The McGraw·Hill Companies

Cataloging-in-Publication Data is on file with the Library of Congress

McGraw-Hill books are available at special quantity discounts to use as premiums and sales promotions, or for use in corporate training programs. For more information, please write to the Director of Special Sales, Professional Publishing, McGraw-Hill, Two Penn Plaza, New York, NY 10121-2298. Or contact your local bookstore.

The IP Multimedia Subsystem (IMS): Session Control and Other Network Operations

1234567890   FGR FGR   01987

ISBN   978-0-07-148853-2
MHID   0-07-148853-7

Sponsoring Editor	**Copy Editor**	**Composition**
Jane Brownlow	Robert Campbell	International Typesetting and Composition
Editorial Supervisor	**Proofreader**	
Jody McKenzie	Divya Kapoor	**Illustration**
		International Typesetting and Composition
Project Manager	**Indexer**	
Vasundhara Sawhney (International Typesetting and Composition)	Wordco Indexing Services	
	Production Supervisor	**Art Director, Cover**
	George Anderson	Jeff Weeks
Acquisitions Coordinator		**Cover Designer**
Jennifer Housh		12E Design

This book is dedicated to the late Marvin Lamond Burlison. Marvin was my Uncle, my friend, and a true Telephone Pioneer. He provided me many insights into the business of telecommunications based on his 41 years of service at Southwestern Bell. He spent his Thursdays during his retirement volunteering at the Telephone Pioneer Store and Museum at Bell One Plaza in Dallas, Texas, and continued serving his community through the charitable work of the Telephone Pioneers as well as his church. Marvin taught us all leadership, humility, and loyalty both to his company and to God. We miss him dearly.

About the Author

Travis Russell has more than 25 years experience in the telecommunications industry. His experience as a field engineer and later as instructor and consultant covers multiple disciplines including switching, signaling, and data networking. He currently has numerous patents pending for solutions and technologies that encompass signaling as well as OSS/BSS. He is the author of *Telecommunications Protocols, The Telecommunications Pocket Reference,* and *Signaling System #7* (currently in its 5th Edition) and co-author of *Cellular Digital Packet Data (CDPD)*.

Mr. Russell is a frequent speaker at colleges and universities, as well as industry conferences and trade shows around the world. He is currently a senior manager with Tekelec, the leading developer of switching and signaling telecommunications solutions.

Contents

Foreword

There are two camps of service providers who make money off of the residential Internet users today. Camp one is made up of the Googles, Yahoos, and AOLs who are often referred to as "over the top" or WEB 2.0 service providers. This camp doesn't need IMS, or at least they haven't shown much interest in IMS. On the other hand, this camp is known for the cool innovative stuff like P2P VoIP (Skype), iTunes, and video PC such as YouTube and more.

Members of the other camp are the legacy, facilities-based wireline, wireless, and cable companies. Regarding IP service, this camp provides Internet broadband access and/or functions as an ISP. The problem with these two services is that they are flat rate and are considered commodities in most markets. For this camp, IMS presents perhaps the only real opportunity to become innovative and profit like the camp one player.

So, what is missing in the Internet architecture that limits facility-based telecom operators from improving their bottom line financially? Or better said, why do they need IMS? First, the Internet was designed for best-effort routing and connectivity. Adopting an IMS architecture over an IP-based infrastructure gives service providers control over what the user is doing. With control, knowing what the user is doing or what they want to do allows the operator to synchronize Quality of Service (QoS) with service provisioning and, equally important, monetizing the transaction. In short, IMS architecture provides a means to provide QoS with compensation.

Second, the camp one portals create value by combining non-telecom services with telecom services or so-called service mash-ups—for example, calling a buddy via VoIP over broadband and delivering a map displayed on a screen that shows where you are calling from. IMS architecture allows a telecom operator to do the same: bundle a non-telecom product with a telephone service (or voice) component.

Finally, today major Tier One telecom operators have many non-interoperable or so-called "smokestack" back-office functions generally referred to as operations and business support systems (OSS/BSS). With IMS architecture, these smokestacks can be reduced to one OSS/BSS, thereby increasing operational efficiency and reducing costs.

So in a nutshell, what's IMS? It's a vast collection of developed functions and interface standards under one umbrella architecture called IMS. The key to its almost universal acceptance by the telecom standards community is that it reuses already established standards such as Session Initiation Protocol (SIP) and DIAMETER.

The bottom line is that if you are a telecom operator employee or supplier addressing the next-generation network architecture, this book is a must read and ready reference. There is no other next-generation architecture for IP networks than IMS.

What makes this book so valuable is that it gives the reader a transitional road map to an IMS architecture as well as a significant clarification of the many functional and interface standards. Eventually everyone involved in the next-generation networks must have an understanding of IMS, and this book is a great starting point.

Dr. Jerry Lucas
President
Telestrategies

Acknowledgments

As always there are many people who make books happen. This project would not have been possible without the help and guidance of my editor, Jane Brownlow. Thanks, Jane, for your persistence and guidance throughout this project.

Of course, pulling all the pieces together from manuscript to figures and illustrations is always a challenge when dealing with us eclectic authors. Jennifer Housh has been invaluable in pulling all these pieces together and creating the final product. Many thanks to Jennifer and the entire production team at McGraw-Hill for their hard work.

I have been fortunate for the last 17+ years to have consistent employment in an industry rife with change and churn. Tekelec has been a great company to my family and me, and we are grateful for the many years of service here. Tekelec has provided the opportunity to learn new technologies and new viewpoints about why these technologies are important to the industry. They have also been very supportive of my books, and for that I am very grateful. Many thanks to the entire Executive Management team at Tekelec for your ongoing support, and of course the many employees and colleagues who make going to work each day a pleasure.

Since this book was written during the evenings and on weekends, my personal life is void of a lot of spare time. There is no element in my life more important than my family. Over the years we have learned how to make the most out of our time together, while still remaining dedicated to writing. Thank you to my beautiful wife for her patience, and to my children Nichole, Laura, and Travis for their understanding and encouragement.

Introduction

Not since the Intelligent Network (IN) and Signaling System #7 (SS7) has there been as much change to the telecommunications networks of the world as we are seeing played out now by the 3GPP and 3GPP2 communities. Much of this change comes from lessons learned, rather than a need for new technology. And as we have seen played out many times before, this radical change relies on existing technologies with a new implementation.

As the wireless world began moving away from traditional network architectures reliant on older time division multiplex (TDM) facilities and into an all-IP infrastructure, it became apparent that there would be many challenges. The most significant of these challenges is supporting the many different functions required in a wireless network to support capabilities such as roaming.

It is not the simple function of roaming itself that presents the challenge. It is the challenge of making sure a subscriber using your network is authorized to use wireless services, has an active account with a partner network, and is authorized to use the services that subscriber is requesting. It is sharing this information and authenticating the subscriber that presents the challenge.

It is also the challenge of preventing unauthorized access from hackers and fraudsters. The GSM community has learned many lessons already on this front and has been successful in eliminating many forms of fraud today through new procedures and technologies in the network. For example, to prevent eavesdropping on the air interface (a common practice in the early days of GSM used to obtain the subscriber's identity), GSM operators implemented encryption over the air interface. This prevented eavesdropping and was responsible for all but eliminating cloning of cell phones.

Wireless service providers have built a model that not only protects them from unauthorized access to their services, but allows them to share certain details about their subscribers as they roam into other networks. And they can do so with confidence that they have a fair amount of security. They have achieved this by implementing wireless technologies in such a way as to eliminate opportunities for fraud.

So when IP is introduced into a traditional telecommunications network, it represents many challenges for the operator. Certainly when wireless providers began deploying IP in their networks, they found issues with making it work as their networks operate today. They found there was a lack of implementation standards that defined how networks were to communicate with one another, and no standards to support authentication in an IP world.

Why is all this necessary? Because traditional service providers earn their revenues through services provided to their subscribers. This is a significant departure from the Internet business model, which operates as a free enterprise, with companies earning revenues through advertising and value-added services rather than access. The Internet companies of the world in fact do not worry about access, because the telephone companies and Internet service providers (ISPs) provide this as part of their service offerings.

What happens, though, when the very telephone companies providing access to the Internet also attempt to provide their services and applications in an Internet model? The mechanisms required by telephone companies to bill for their many services, and to ensure that the network is secure and operating at its highest efficiency, are no longer available.

For example, on the Internet, one can send an e-mail anonymously, using one of many different anonymous servers on the Internet. And the e-mail service is provided at no charge (look at Yahoo, for example). Now imagine a telephone company using the same model, with no means of recovering its investments in the e-mail platforms, and no means of tracing and validating the originator of the e-mail. In fact, imagine not even being able to determine if the subscriber sending or receiving the e-mail is even authorized to use e-mail services.

Telecommunication companies definitely have a different business model, and it's this difference that requires major changes in the architecture and the implementation of IP architectures prior to offering network-wide services on IP. This is what the 3GPP set out to change.

The 3GPP community has learned from many earlier mistakes of implementing wireless technologies. Just look at the evolution of GSM and you can see the numerous improvements made over the years making GSM more robust and more secure. As issues arose (such as handset cloning), the GSM community devised new tactics to solve the problem (such as encryption over the air interface). So why IMS?

First, providing multiple services with multimedia requires management of the Quality of Service (QoS). This means being able to assign new facilities to support bandwidth needed when a new media type is added, for example. Providing this level of QoS management is best suited for a model such as the IMS. However, there is another very key function that the IMS brings that traditional IP networks could not support.

As the 3GPP began looking at ways to implement IP in their networks, they found a need to standardize first of all on one common protocol for call and session control. This was missing from any standards and is the reason why today you see many companies with multiple versions of VoIP protocols in their networks. It became a contest to see which vendor could sell their products based on proprietary versions of multiple protocol standards in an effort to lock operators into their proprietary network model (which means buying all of their network elements from one vendor).

This is what has led to the evolution of IMS. The IMS architecture defines one common protocol standard to be used network-wide for all sessions within the wireless network. This eliminates the multiple protocol nightmare that exists in the VoIP world now. Operators no longer need to worry about purchasing network elements that support the same version of H.248 as their media gateway controllers (MGCs). But it goes further than that.

The 3GPP has identified specific functions that must be supported within their networks to ensure security and to support authentication and the sharing of subscriber data with roaming partners. They have added some new functions to support this important aspect of the network: call session control (or the Call Session Control Function, or CSCF). The CSCF is the element in the IMS that brings session control back to the core where it can be managed and scaled.

There are many vendors who will promote the idea that the CSCF is best supported at the edge of the network, residing on the access platforms such as media gateway controllers (MGCs), but this model has already been tried. It was this very model that led to the implementation of SS7 and the IN. The concept of an IN was a brilliant concept, but it failed because the cost to implement an IN was too prohibitive. Software required on every switch in the network for accessing the IN is what raised the cost of implementation of an IN, although the benefits were clear. It was wireless that demonstrated the value of core functions such as call control through the support of features and capabilities such as roaming.

IMS builds on this theme, providing the additional benefit of controlling all communications, not just voice and data. Supporting messaging, e-mail, and file downloads under one common control provides many economies of scale to service providers and affords them the ability to provide new services (such as presence) that are not feasible using other models.

The IMS is yet another level of evolution within the GSM strategy, yet it provides many benefits to cable and wireline providers as well. IMS is not so much a new technology as it is an implementation standard for existing technologies. SIP for example is not new but is already implemented in numerous VoIP networks worldwide. The IMS brings SIP to a core function where everything can be controlled and managed more efficiently. It also provides security for wireline providers that does not seem to be there in traditional VoIP implementations.

So what is the business case for IMS? This is the most burning question operators around the world are asking right now. Those that have been working with the 3GPP understand the business case, because they understand the value that IMS brings to their business. The IMS does not represent new services, as many marketing departments will tout. True, the IMS is a service enabler, but this makes a tough business case.

The IMS brings a much more effective and secure means of providing multimedia services to your subscriber base, utilizing an all-IP core. It's about economy rather than revenue generation. The best way to think of economy is to consider the margins your services bring today. If you can reduce the cost of service implementation and make it more secure (eliminating or reducing revenue loss), your margins increase, therefore making IMS a better business case for delivering new services.

All of that being said, IMS is not something that happens overnight. One only needs to look back through history to understand the dynamics of introducing new technologies into a service provider network. SS7 began development as early as 1964, yet it was not implemented in the U.S. until the mid-1980s. Likewise, ATM was originally documented in 1968 yet not implemented in the core network until the 1990s (in large-scale deployments). There are many reasons it takes so long for new technology to make its way into the network.

There are significant investments that have been made in existing networks. These investments do not get replaced until they have been fully depreciated from an operator's books. This, of course, always presents a challenge for vendors who must develop and then sell product to pay for the development of new network entities. We are right now in the phase where operators begin planning and deploying segments of IMS in their networks, but large-scale deployments may take many years to come.

This means that SS7 will continue to exist in the network for many years to come. The IMS networks being deployed today are being deployed next to the existing SS7 networks, and are interworking with these networks through the procedures defined by the 3GPP (and explained in this book). So SS7 does not go away tomorrow but will continue to grow, in fact, in many regions. Slowly, segments of IMS deployments will begin to take over SS7 functions, until the SS7 network is completely replaced by the IMS.

This book provides an interpretation of the 3GPP standards for IMS. This is not the view of any one vendor, nor is it the view of the 3GPP. Rather, this an interpretation of the 3GPP specifications that define the IMS and its functions. What makes this book different is the insight into the Intelligent Network, as well as years of background and experience in data and voice network design.

The intent of this book is not to provide a how-to guide for IMS implementation, but rather to offer an explanation to why the various functions within the IMS do what they do, and what distinguishes this architecture from other implementations of IP transport.

It is hopeful that through this book readers will gain a better understanding as to why IMS was developed to begin with, and what value an IMS architecture brings. Since the IMS standards themselves continue to evolve, it is likely there will be updates to this book over the next decade.

The SIP protocol is not covered in this book but will be the subject of my next book, *Session Initiation Protocol (SIP): Controlling Convergent Networks* (McGraw-Hill, forthcoming). For a complete view from PSTN to the IMS, don't forget my earlier book *Signaling System #7* (McGraw-Hill, 2002), now in its fifth edition. Collectively, these three books provide an end-to-end view of the signaling and call control within both traditional and next-generation networks for both wireline and wireless.

1

Architecture of the IMS

There are many different viewpoints and definitions for the IP Multimedia Subsystem (IMS). As with any new technology introduction, as vendors work diligently to develop products for the new technologies, there is also the period of time when there is a lot of marketing effort to try to promote the implementation of the technology.

Today we see this with IMS. While many are promoting the IMS as a simple call control architecture easily implemented at the edge of the network, the 3rd Generation Partnership Project (3GPP) has defined a much more robust approach.

There are many who believe the IMS is all about services. Yet others believe it is all about control. In reality, it is a little bit about both, but the primary purpose of the IMS infrastructure is to provide session control at the core of the network while enabling other support needed to provide those services, regardless of media type.

IMS Concepts

When people talk about IMS services, they talk about multimedia sessions consisting of video, audio, and voice. They talk about mixing all of these media types in a single session, and, of course, being able to use these services while roaming the network. This is absolutely true, and is what makes the IMS different from the telecommunications networks of today.

Today's networks support voice, but other media must be supported through network overlays. The operations support systems (OSSs) and the back office support systems (BSSs) needed to provide billing, mediation, provisioning, diagnostics, and all of the other support mechanisms needed are piecemeal today, consisting of many different disparate systems rather than one architecture supporting all media types.

The most common argument that comes up during seminars and conferences is that multimedia services can be supported today, and in fact there are many operators who are doing just that. While this is true, when you speak to those operators about how they are delivering these services, you quickly learn that while it is possible to support

a multimedia service using today's IP networks, it is another thing to be able to accurately bill for that service, and to monitor the QoS for that service. Troubleshooting
these networks is troublesome, and providing an accurate audit of content downloads
is challenging.

Another major issue that was one of the drivers for the IMS is interoperability.
During the early implementation of Voice over IP (VoIP), interoperability between different vendors was difficult. This is why the 3GPP community set about defining a
standard implementation for IP services. The IMS is not about some new technology to
displace all current technology, but rather another phase in the evolution of wireless
networks.

We are indeed seeing an evolution of networks. The current network is well suited for
providing voice, and the technology can be modified to support other media types, but
more than just transmission technologies are needed to support multimedia services.
Session control, security, and charging are all important aspects for service delivery.
Providing a core infrastructure that can support these services is important.

This is what IMS is all about. There is no argument that TCP/IP can be used today to
deliver multimedia services. Many of us are using these services today through sources
such as AOL and Google. However, how those services are managed, how they are made
secure and how billing is implemented constitute another challenge.

The IMS is not about services. The IMS is about providing access to all services regardless of the media type, using a common control architecture that works well for all
media. This means that one common control plane is used for video, voice, data, messaging, and any other media format needed. What's even more important is that the control
plane provided through IMS does not need modification to support a new media type.
It does not need a different technology for each media type. Everything is controlled
through one common session control protocol: the Session Initiation Protocol (SIP).

Another fundamental difference with IMS is security. When one looks at today's VoIP
implementations, security is sometimes part of the implementation, but it is not very ag-
gressive. The 3GPP has defined new security functions as a part of the IMS model, and
a base function of the call session control function that establishes the core of the IMS.

The 3GPP has really defined how to approach implementing an all-IP network with
robust security, charging, and session control at the core of the network rather than in
many different boxes, all interconnecting with one another.

What is really interesting about the IMS model is how everything is controlled using
one common signaling method. Signaling is what makes connections happen in any
network. It is signaling that allows network entities to communicate with one another
as well as application servers regarding the service to be delivered.

In the Public Switched Telephone Network (PSTN) today, Signaling System #7 (SS7)
is the signaling method used to connect facilities in the telephone network. SS7 connects different telephone companies networks with one another, and connects all of the
switches within the network with one another.

In the IMS, SS7 is replaced with a new signaling method: SIP. SIP is used to control
everything in the network, allowing all network entities to communicate with one another regarding service delivery network-wide.

So when we begin talking about the IMS architecture, we will be looking into how all of the media types and the various entities are controlled through SIP. However, SIP alone is not enough. There are network functions that must be developed that work with SIP to provide a universal control mechanism for all forms of subscriber communications. The IMS defines these functions as well, and we will be looking at these functions in detail.

So don't think of the IMS as a means of delivering killer new applications, and don't look for a "killer app" in IMS. The "killer app" is something far simpler than you may think. It's simply providing an infrastructure that supports how subscribers communicate with one another—while shopping, while driving, while boating. By sending text, pictures, audio clips, music files, video streams, and voice. Think of IMS as the man behind the curtain, providing control over all of these communications.

The Intelligent Network and Signaling System #7

Today's networks were designed primarily for voice communications. Sometime in the early 1960s it was determined that the then-current method of call control was not only inefficient, it was highly susceptible to fraud and did not support the delivery of applications such as Freephone (800 in the U.S.) and Calling Name.

Another fundamental problem with telephone networks was the requirement for switches to communicate with one another to exchange dialed digits and other call data. Prior to the digitization of the world's networks, most of this communication was done over telephone circuits using audible tones.

However, through digital communications, the industry was about to expand the communications between the network entities to include much more. This is when the Intelligent Network (IN) was conceived.

The telephone industry set out to create a new architecture that would provide more call control central in the network, and support the delivery of new services. The outcome of this work was what we refer to today as the Intelligent Network. The IN introduced a new layer into the network, specifically designed for controlling all telephone calls that required a trunk to connect to another switch. In very basic terms, the IN is a data network that allows switches and databases to communicate with one another autonomously regarding voice services, telephone call by telephone call.

Another major advantage of the IN was the ability to put more intelligence into the core of the network. This was important, because subscribers wanted more than just voice communications. Consider how delivery of a Calling Name would be accomplished without the IN.

Calling Name delivery requires a database that is capable of matching the calling party telephone number with a name. If we were to try to provide this service without the IN, it would require this database to be resident within every single switch in the network. Databases are not cheap, so this would require a major investment on the part of the service provider.

Cost aside, the simple maintenance of such a database would also be a major undertaking. Any time a single change was to be made to the database, the change would have to be replicated at every single switch.

However, if the database could be placed in the core of the network, the switches could then connect to the single database and all take advantage of a single resource. It would also make administration much easier, as the database changes would not have to be replicated many times. This of course would require communications between the switches and the database.

To facilitate these communications, a new control protocol was developed called Signaling System #7 (SS7). SS7 (or C7 as it is often referred to in countries outside the U.S.) is used by switches to send information about a telephone call to another switch prior to connecting the trunk and routing the call over that trunk. It allows switches to communicate to one another the requirements for the telephone call, and it also allows specific services to be identified prior to routing the call.

The IN and SS7 also support databases in the core network for providing additional instructions for treatment of a call. These databases have become especially important in wireless networks that rely heavily on the SS7 and the IN. For example, the database used by wireless networks to store information about their subscribers is referred to as the Home Location Register (HLR). The HLR is a central function in any wireless network and is accessed through the IN using the SS7 protocol.

Number Portability is another critical function that relies on the IN and SS7. Number Portability would be impossible without the IN and SS7. The list goes on and on, and includes many features and applications we take for granted today. Without the ability for switches to communicate and query these databases, we would not be able to provide many services we enjoy today.

However, the IN has run out of gas. The IN and SS7 were designed specifically for voice and data communications; they are not equipped for other media types. The entire protocol would require modification to be able to support IP networks and multimedia services. Even then, the IN and SS7 are not suitable for IP-based multimedia, and therefore the entire infrastructure for the network core has been redefined. The very communications and the procedures defined for call treatment had to be recreated to support multimedia over an IP network.

And thus the Session Initiation Protocol (SIP) was born. SIP is like SS7 in some aspects, but provides much more than SS7 ever could. It was used first for Voice over IP (VoIP) networks as a control protocol and has quickly gained in popularity. It was adapted by the 3GPP for use within the IMS core as the control protocol of choice, and it forms the basis of all things IMS.

The principle difference between SS7 and SIP lies in the ability to support various media types. The structure of the protocol is fundamentally different; that is why it can support all media types. Consideration was also given to how services are charged within an IP network, with SIP playing a major role.

I should note here, however, that the first versions of SIP developed for VoIP are not robust enough to support the IMS. The 3GPP has added many extensions to SIP to make this a much more robust protocol, and to support the many functions defined for security and authorization.

So SIP becomes the new call session control protocol for all things in the IMS network. As additional processes and procedures are developed and defined, SIP will continue to play a focal role. For lawful intercept, for example, standards call upon SIP to provide the needed data regarding a call or session. This is a departure from previous standards that required switches and network entities to capture call details and provide this data through a different interface, dedicated to the function of lawful intercept.

For wireless environments, many aspects of SIP had to be modified. For example, in the GSM network, a subscriber device must register with the network prior to receiving service. This simple act of registration is how the network maintains the location of a device for call delivery. It is also how subscribers are authenticated for security purposes, with the device itself exchanging credentials with the network during the registration process.

This concept of registration was carried over from the GSM environment into the IMS domain, and will be evident as you read through the other chapters in this book. This is powerful for any VoIP implementation because today many VoIP networks are highly susceptible to fraud. This is partly because of the lack of enforced procedures such as authentication.

SIP does have its problems however. As with any new technology there is room for improvement. The same holds true for the IMS architecture. Certainly there is room for improvement; however, the benefits far outweigh the shortcomings.

The primary benefit in this author's mind is security. When looking at VoIP issues, many of those issues center around fraud and access breaches. This is because the version of SIP defined for this domain is not very robust, since it was developed in the spirit of the public Internet.

Before going on to discuss the various network entities in more detail, we will first look at current network infrastructures so that you can better understand the differences as well as the concepts of telecommunication networks. The IN architecture is shown in Figure 1.1. Note that the database functions are more central in the network. Their role is not defined as part of the IN architecture, since their role differs depending on the service they are providing. The databases in the network are used to deliver the applications such as Calling Name and Number Portability.

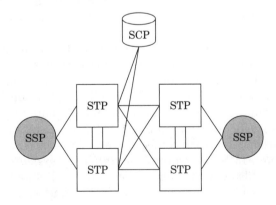

Figure 1.1   The Intelligent Network (wireline implementation)

The central core function in the IN is the *signal transfer point (STP)*. It is the STP that provides the call control within the IN model. Without the STP, this network model would be an absolute mess with every switch in the network interconnecting to one another (a mesh network). The mesh network is simply not scalable, as has already been proven many times before in legacy networks.

The STP in the IN is first a router for all signaling messages. All switches and other network entities connect to an STP, which has responsibility for routing the message to other entities for processing. When signaling is sent to another network, the STP provides a gateway function. Part of this gateway function includes gateway screening, which is used to control the access allowed by other operator networks.

So the STP is both router and firewall for the signaling network. It determines (through operator provisioning, of course) what signaling messages are allowed into the network, and how those messages are to be routed to their destination.

The STP also provides a routing function to the various network databases. If implemented properly, all database addresses are provisioned in the routing tables of the STP rather than the switches. All database queries are then routed to the STP in the core network, which then has the responsibility of routing to the most available database *(service control point, or SCP)*. By using the STP to route these queries, you add an additional level of security within the network, and eliminate the need to provision SCP addresses in every node throughout the network. The SCP is where the service such as HLR, Number Portability, or Calling Name resides.

This becomes important because as databases grow, a need arises to add additional elements into the network. If an operator is providing a Number Portability function, and it has to add another database to support the service, the routing tables would require changing in every switch accessing the database. STPs and a function known as Global Title Translation (GTT) eliminate this requirement.

This is a very simple control network, and not all that dissimilar to the IMS. When you examine the functions within the IMS, think of a wireless network and its entities (as identified in Figure 1.2) and you will have a clearer idea of how the IMS works.

In the wireless implementation, the messaging between databases and switches increases significantly. These databases provide an important function to the wireless network such as location, registration, and authentication/authorization. These concepts have been extended for use in the IMS, which comes as little surprise given the architects of IMS are also the architects of the GSM architecture.

The communications in the wireless network pass between the switches and the base stations themselves. The base stations communicate with the switches using the SS7 protocol as a transport. The Mobile Application Part (MAP) carries the real signaling information that is relevant to mobile communications. MAP supports all of the wireless-specific functions such as the authentication and registration processes.

So in a wireless network today, there is SS7 communication between the *mobile switching centers (MSCs)* and the *base station subsystems (BSSs)*, as well as from MSC to MSC. SS7 is also used to communicate to the Public Switched Telephone Network (PSTN) when a call is routed between wireline and wireless domains.

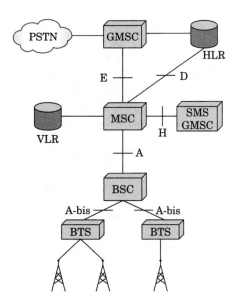

Figure 1.2   The Intelligent Network (wireless implementation)

The switches and the base stations communicate the location of a subscriber device as they roam from cell site to cell site, providing the location data to the Home Location Register (HLR) and the Visited Location Register (VLR). The HLR is a more static database providing the subscription data as well as identifying the serving MSC. The VLR is a more dynamic database identifying the cell site serving a subscriber device.

Something else IMS inherits from wireless networks is the concept of home and visited networks. In the wireless network, subscribers are assigned to a "home" network, which becomes the location of their subscriber data (provisioned in the network database function HLR). Whenever a subscriber roams outside of their network, they are considered as "visitor." Their location is tracked through the same databases in the home network to allow the network to track and authorize access to specific services such as messaging.

When Voice over IP (VoIP) was introduced, wireless operators struggled with making VoIP implementations support their wireless networks. The updating of the subscriber data and other myriad functions defined in the wireless domain was not supported in VoIP protocols, and SIP did not support multimedia communications such as messaging very efficiently.

There were also many issues with interoperability experienced by the wireline operators implementing VoIP, which had to be addressed prior to implementation in the wireless domain. These were some of the drivers that led the 3GPP to define a new implementation standard called the IMS. The IMS is really about implementing VoIP in a wireless network, supporting the requirements of a wireless operator.

Certainly IMS makes a great model for wireless, which is what it was originally designed for as part of the evolution of wireless networks. However, as the IMS evolved and further developed, wireline operators saw the value for use in their own networks, and IMS quickly gained popularity for wireline. Figure 1.3 illustrates the migration from 2G Wireless, to 2.5G, then 3G, on to the IMS and its role.

The evolution of wireless began with simple voice services using the IN and SS7 to provide call control as well as service control. Since the services were limited to voice, this was fairly easy to accomplish; however, without IN and SS7, mobility would have never been possible. The IN and SS7 provide the core call control functions throughout the wireless domain. Remember it is SS7 and the MAP protocols that allow the wireless switches and base stations to communicate with each other, providing registration and location updates for each subscriber device.

When data services were added to this network, there came a need to offload the data from the MSC onto a data network. Referred to as 2.5G, this consists of the General Packet Radio Service (GPRS) network. The purpose of the GPRS network is to provide the functions necessary to support packet services.

In other words, instead of sending the data packets through the base stations, to the MSC, and then offloading the packet data to another network, the data is sent straight from the base stations to the packet network (GPRS).

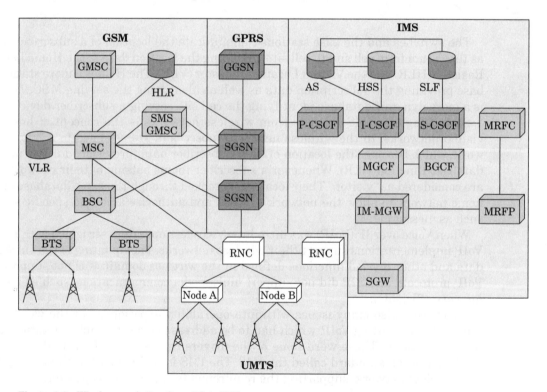

Figure 1.3 Wireless evolution from 2G to IMS

The GPRS then interconnects the wireless domain with the Internet and other packet networks, eliminating the need entirely for sending the packet data through the MSC. The Serving GPRS Support Node (SGSN) and the Gateway GPRS Support Node (GGSN) both support the role of controlling the data transfer to other packet networks.

The GPRS protocol is obviously different than SS7, which does not support packet services. The GPRS protocol and its associated interfaces support connectivity to the various SGSN/GGSNs within a wireless network, and the control of those connections (referred to as Packet Data Protocol [PDP] *contexts*).

Yet GPRS answers the needs for supporting packet service in the core, and not the air interface. With the addition of messaging and video services came the need for more bandwidth at the air interface, and thus UMTS was created. The UMTS network only covers the air interface consisting of new radio functionality interfacing back into the traditional GSM network as well as the GPRS packet network.

There are other similar technologies that enable bandwidth at the air interface in support of broadband services. Code Division Multiple Access (CDMA) 2000 is one example, as is Enhanced GPRS. Ultimately, SIP will become the standard across-the-air interface as well, once mobile devices have been SIP-enabled.

The IMS extends all of these capabilities using SIP as the primary session control once the session reaches the IMS domain. Eventually, handsets will be equipped with SIP capability and will perform as SIP user agents. Once these phones are available, the legacy wireless network is no longer needed. Handsets will be able to interface via UMTS and other broadband radio interfaces directly into the IMS via the P-CSCF function.

One thing is certain about IMS: while it is being developed and implemented to replace the IN and SS7, this replacement does not happen overnight. It will be quite some time before all of the SS7 network is replaced, and there will be many parts of this network that are never replaced.

This is evident when one looks at how long it has taken other, similar technologies to be fully developed and implemented. For example, the first discussions and papers on Asynchronous Transfer Mode (ATM) date back to 1969 (this author has some of the early papers from the Bell Labs).

Yet it took us nearly 20 years before the technology was really ready for operators to begin implementing in their networks. The same is true regarding SS7 and the IN. The development of SS7 began in the mid-1960s, yet it was not really implemented in the U.S. until the mid-1980s.

Even then it was only implemented to support 8*xx* services and not for call control. It was not until later that operators in the U.S. began implementing SS7 for call control (using the ISDN User Part, or ISUP, protocol).

The migration to SS7 has been even slower in Europe and Asia. Most of the major operators in Europe, for example, have implemented SS7 only for call control between switches, but the role of the STP as a hub within the core of the IN was never fully appreciated until wireless networks began being deployed. In wireline networks it has been only recently that operators have begun widespread expansion of their SS7 networks to support Number Portability applications.

So considering what we have seen in this industry historically, we can safely say that IMS implementations have begun already, but it will be a slow implementation. Indeed operators and vendors alike are estimating IMS will take many years to fully mature and replace the existing SS7 infrastructure.

This means that the legacy networks and the IMS networks will have to interoperate with one another. There are a number of functions that have been defined for interoperating between the legacy network and the IMS network, which are explained in more detail in Chapter 2.

Entities in an IMS Core Network

As stated already, the main focus of the IMS is session control. When one looks carefully at the 3GPP standards for IMS, you will not see any switching functions as part of the architecture. Switches, and network nodes used for access, are not part of the IMS model. They are used simply to access the IMS, but the IMS does not care how a subscriber enters the network.

To the IMS, the access network is transparent. There are gateway functions that convert the legacy signaling (mostly SS7) to SIP prior to interfacing to the IMS. This is important to understand, because as one begins to look at possible implementation solutions, it can become confusing as to how the architecture is different than what we have today in standard legacy switch-based networks.

The access network therefore is any wireline, wireless, or VoIP network. At some point there is an entry point where IMS procedures begin, as outlined in the following sections. The roles of each of the entities defined in the IMS model are provided in the sections that follow.

Call Session Control Function (CSCF)

The heart and soul of the IMS lies within this function. Call session control is the primary function of the network core. The function of call session control is actually distributed throughout the network to make the network more efficient and scalable, rather than trying to support this function through one huge centralized platform.

At the same time, one probably does not want to distribute this function at every switch. This then becomes unmanageable, and as the network evolves and grows, it becomes unmanageable as well. Cost is what prevented applications from being implemented at the switch level, and it is cost and the inability to manage and maintain very large IP networks at the switch level.

There is no rule that the various call session control functions must be distributed. Indeed there will be some implementations (especially in the early deployments) where this will be the case. There will probably be many instances where all three CSCFs will be implemented in the same network entity. But when the network begins to expand and grow, it begins to make more sense to separate the various control functions into separate, discrete entities.

There are three entities defined that are responsible for call session control:

- **P-CSCF**   The Proxy Call Session Control Function
- **I-CSCF**   The Interrogating Call Session Control Function (I-CSCF)
- **S-CSCF**   The Serving Call Session Control Function (S-CSCF)

The difference between these entities (shown in Figure 1.4) lies in their individual purpose and procedures they perform. Each entity acts as a stateful proxy (although they can also be stateless); therefore, each entity maintains details about all sessions in progress, as well as registration status of the subscriber device.

As I will discuss in later sections and chapters, dividing these three functions into separate entities also makes sense for security purposes. There are specific security functions that have been defined for each of the CSCFs that are best suited for a distributed architecture.

Proxy CSCF (P-CSCF)   The first access point into the IMS is the P-CSCF. The P-CSCF acts as the access point to the SIP domain from a session control perspective. Bearer traffic is not passed through this portion of the IMS, as this is a signaling and control network. The bearer traffic is passed through IP and uses the various access methods for transport.

What passes through the P-CSCF is SIP. The first communications are to register the location of a device; location being the IP address of the device in its present location. As the device communicates with other devices, it must first establish a session with the device. These session establishments are also made first through the P-CSCF.

When a subscriber device is first activated, it will be assigned an IP address by the serving network. Once the device has been assigned an IP address, it will search out the local P-CSCF (or whatever P-CSCF has been assigned to serve this part of the network). The P-CSCF, like all IMS entities, has an address in the form of a SIP Universal Resource Identifier (URI) (making it easier to route messages to the proper P-CSCF).

Once the device is powered on, it sends its IP address to the Home Subscriber Server (HSS) and the S-CSCF using a registration process. The P-CSCF plays an important role in the registration process, as you will learn later. The first role, however, is to identify the home network from the subscriber's domain (found in the URI of the subscriber address). The domain name of the home network is of course resolved into an IP address using the Domain Name Server (DNS) function of the network.

The DNS identifies the address of the I-CSCF to be used to access the home network. The role of the I-CSCF is described in the next section, but for the sake of this discussion, the I-CSCF provides the gateway access into any network.

Figure 1.4   The IMS core and the CSCF

It is up to the P-CSCF to determine how to route any SIP messages received by the subscriber device. For example, when the P-CSCF receives an *INVITE,* it must decide where the *INVITE* is to be sent. The P-CSCF acts as the access point into the IMS but not into individual networks (at least in terms of SIP messaging). The I-CSCF provides further routing to the proper S-CSCF according to registration procedures.

Like all of the CSCF entities within the IMS, the P-CSCF generates CDRs for all sessions that pass through it. The P-CSCF also adds headers to request and response messages before forwarding them onward to the next CSCF. For example, the *access-network-charging-info* parameter is added to the *P-CHARGING-VECTOR* header prior to sending the request to the S-CSCF. It is the P-CSCF that generates the *IMS Charging Identifier (ICID)* used in the charging procedures for correlation of sessions and charges, since the P-CSCF is the entry point into the IMS.

These headers are added so that entities can exchange charging data within the SIP message itself, without having to support yet another interface for charging. However, there is also a charging interface and separate charging overlay that supports the DIAMETER protocol. This is described in greater detail in Chapter 6.

Headers used for charging are not shared with other networks; the P-CSCF only forwards these headers to functions within the same network. This prevents other service providers from learning about subscribers and their service usage. This is yet another function of the P-CSCF that is based on policy decisions configured by the operator.

From a security perspective, the P-CSCF is critical in preventing unauthorized access to the network. Since the P-CSCF is the entry point into the IMS, the P-CSCF can be used to screen access by any device. However, the P-CSCF does not enforce authentication within the IMS. The S-CSCF is responsible for challenging devices when they attempt to establish a session without being registered, or when they are attempting registration.

A Policy Decision Function (PDF) can be resident in the P-CSCF and used to determine how to react to specific scenarios. The PDF allows operators to establish rules to be applied for access to the network. The PDF controls the Policy Enforcement Function (PEF) in the bearer network. This allows operators to control the flow of packets at the bearer level according to destination and origination addresses and permissions.

The P-CSCF can also check the routing to verify that the routing received in the SIP request/response (as identified in the *ROUTE* header) is the same routing that was identified when the device registered in the network. If the routing headers do not contain an address matching the addresses saved by the P-CSCF during the registration process, then the routing is changed by the P-CSCF in accordance with the addresses captured in the P-CSCF. This is to prevent hijacking and other scenarios where a hacker may capture a SIP message and use it as a duplicate to obtain services from another portion of the network. This would result in the message being received by the P-CSCF with a different set of addresses in the routing headers.

The P-CSCF is able to provide this function because when an entity registers in the network, the P-CSCF saves all of the addresses provided in the routing headers. While the addresses do not have to be in order, they do have to be present as addresses that were identified during registration. This function prevents some types of attacks and is another form of security provided by the IMS core. Other information saved by the P-CSCF includes the device address (IP address) and the public and private user identities.

If a device loses its connection in the IP network, the P-CSCF is notified and releases all sessions within the IMS by sending a CANCEL to any entities that are part of the session. Since the P-CSCF is the first point of contact for all devices in the IMS, the P-CSCF has knowledge of all sessions created through it. The P-CSCF is stateful, so it also has knowledge about the state of each session.

There are times when the P-CSCF can function as a SIP User Agent (UA) when subscribing to event notification, or other non-session-related subscriptions. This means that there can be instances where the P-CSCF will initiate requests toward an Application Server (AS) or S-CSCF.

So one should think of the P-CSCF as the gateway into the IMS network itself. It has the responsibility to ensure that the device accessing the network has been registered and is allowed access into the IMS, but it does not enforce authentication and authorization. The P-CSCF also determines how to route the SIP messages received to the proper networks and network entities, allowing operators to maintain some level of secrecy regarding their own individual networks, and eliminating the need to create routing tables at every single node (the mesh network topology).

Interrogating CSCF (I-CSCF)   While the P-CSCF is the entry point into the IMS, the I-CSCF serves as the gateway into each individual IMS network. It is the I-CSCF that determines whether or not access is granted to other networks forwarding SIP messages to the operator. For this reason, the I-CSCF can be used to hide network details from other operators, determining routing within the trusted domain. The I-CSCF helps to protect the S-CSCF and the HSS from unauthorized access by other networks.

When the S-CSCF is forwarding a request or a response to another network, the message is forwarded to the I-CSCF, which in turn forwards it to the destination network. The I-CSCF in the other network then has the responsibility of identifying the location of the user being addressed (possibly through the Subscriber Locator Function).

The location means identifying the S-CSCF assigned to the user, as well as the HSS where the subscription data is stored. The SLF can provide this information to the I-CSCF. This becomes an important role within any IMS helping to secure access and further protect the identities (routing addresses) of the S-CSCF and HSS resources.

This is somewhat different than the gateway function provided by the P-CSCF. The P-CSCF acts as a gateway to non-IMS networks and the access point to reach the IMS, while the I-CSCF acts as a gateway between two IMS networks.

Another important function of the I-CSCF is the assignment of the S-CSCF. The S-CSCF is assigned according to capability or service provider policy. There are two options available for implementation. One approach is to assign the S-CSCF according to the services that need to be supported. For example, if a video conference is being set up, the S-CSCF providing access to video resources is assigned. This approach may be favorable to service providers looking to distribute media support throughout the network, or in specific portions of the network.

This model works well in data networks where services and platforms are located according to the media to be supported. In an ASP model, for example, the video server may be located within a data center, along with an S-CSCF supporting video services. All subscriptions for those video services would then be assigned to the S-CSCF in that location. The decision of how to assign the S-CSCF then becomes based on the subscriptions themselves, rather than where the subscription is located.

The other approach is to assign each S-CSCF according to geography. This is the traditional telecom approach used in today's legacy networks with switches. This model works well for traditional telecom operators, and in some networks it may make the most sense. The S-CSCF is then assigned according to the location of the I-CSCF that receives the request/response. The S-CSCF assignment is stored in the Home Subscriber Server (HSS) for future reference.

For wireless and wireline operators, this model probably makes the most sense, given that this is the model they are accustomed to today. Subscribers are assigned to resources according to their home locations, and all services are supported network-wide rather than in network segments.

The S-CSCF is assigned by the I-CSCF when a subscriber registers in the network. The I-CSCF then stores this information (along with routing information) throughout the lifetime of the registration. When the I-CSCF receives subsequent requests/responses, it forwards them to the S-CSCF assigned on the basis of this information.

A registration lasts as long as the subscriber device remains in the service area. For example, in a wireless network, as long as the subscriber device continues to receive service from the same cell site, the registration remains alive. However, as soon as the subscriber moves to another cell site, the registration is changed (because the cell site address is changed).

In a wireline model, as long as the subscriber device maintains the same IP address, the registration remains alive. When the subscriber is assigned a new IP address, then the subscriber device must register its new IP address.

Of course, there is a timer associated with all registrations. Each registration has an expiration, and once the expiration is reached, the registration will be terminated by the S-CSCF and HSS. This prevents a subscriber device from registering with the network and then disconnecting without canceling the registration. That would of course provide an opportunity for a hacker to come along and take advantage of the registration.

To prevent exposure of network topology (the number of S-CSCFs, addresses of internal entities, etc.) to non-trusted networks, the I-CSCF provides a function known as topology hiding. Topology hiding entails removal of certain SIP headers to prevent other networks from learning about the network.

For example, the *ROUTE* header contains addresses of all the IMS entities that were used to route the message. These could be used to identify the number of hops to the S-CSCF, as well as other "proprietary" network information that the operator may not want to share with another network.

If a hacker were able to capture all of the SIP messages using a protocol analyzer (or other method of sniffing on the network), the hacker could also use this information to determine round trip time in the network and calculate other parameters that could then be used in a denial of service attack. Security measures are described in more detail in Chapter 5.

The I-CSCF could be considered as your firewall into your IMS network, routing between other networks, and preventing other networks from accessing your network. The I-CSCF function therefore plays an important role in security for IMS.

Serving CSCF (S-CSCF)   The S-CSCF serves as the core to the IMS. It controls all aspects of a subscriber's service, maintaining status on every session he or she initiates. Remember that a session is anything that the device wishes to do, including e-mail, instant messaging, sending pictures to a mailing list, and voice. The S-CSCF controls messaging and delivery of content. It provides the status of a subscriber's registration to other applications (application servers) and maintains control over these services as long as the device is registered.

From a SIP perspective, the S-CSCF is the registrar, responsible for authenticating all subscribers who attempt to register their location with the network. When challenged, the S-CSCF will force the subscriber device to send another *REGISTER* message carrying the proper credentials and authentication keys prior to granting access to services.

This is one function that is not common in today's VoIP implementations. Too often softswitches (which is where the call control is managed today) do not have robust security functions, and they do not challenge every subscriber device when accessed. This is one of the areas where the 3GPP specifications have improved on VoIP security.

The S-CSCF saves the following information about a registered device after registration:

- HSS address
- User profile
- P-CSCF address (the entry point during registration)
- P-CSCF domain (in the event the device entered through another network)
- Public user identity
- Private user identity
- Device IP address

Duplicating this function throughout the network in a mesh configuration would be extremely difficult, not to mention inefficient. Using a core function for security and access control is the spirit of the IMS and has already been proven as a winning proposition by wireless providers.

Placing these functions at the edge of the network (as in Session Border Control) also does not make sense, because many times attacks are made from within the network. Not to mention the issue of authentication at the network's borders. The border devices would still need to know the authentication and cipher keys for every subscriber, and they would still have to access some centralized function for this information.

The S-CSCF also has the responsibility for enabling services by providing access to various Application Servers (ASs) within the network. For example, if a subscriber device is attempting to connect to a video conference, the S-CSCF would provide the connectivity to the proper AS in accordance with the subscription (as defined within the HSS) and the media requirements defined in the SIP Session Description Protocol (SDP).

This means that the S-CSCF needs to know what services the subscription is allowed access to, and the addresses of the servers providing these services. The S-CSCF accesses the HSS to identify the subscription profile, which includes the service profile. These are explained in more detail in Chapter 3.

The S-CSCF may need to provide forking capability for some services such as conferencing. For example, during a video conference, the subscriber device may wish to invite multiple users to the conference. The S-CSCF forks the *INVITE* using SIP procedures for forking proxies, sending the request to the addressed parties. Forking preferences are usually identified by the device during registration, but if no preferences are provided by the device, the S-CSCF is then able to determine how to route the request.

Converting addresses may also be a role of the S-CSCF (as well as the P-CSCF and the I-CSCF). Since SIP routes are based on SIP URIs, any TEL URIs must be translated or converted to a SIP URI. The same is true in the opposite direction when routing from the IMS to the PSTN. The S-CSCF is responsible for accessing an ENUM/DNS application for converting these addresses into a SIP URI prior to forwarding the request (or response) to its destination.

The ENUM/DNS application can be colocated within the same application server, or it may be a stand-alone ENUM function. The purpose is to enable routing between the legacy networks where E.164 numbers are still used to reach subscribers, while still supporting the concept of the SIP URI.

S-CSCF operates as a stateful proxy and must always maintain the state of all registrations and sessions under its control. Since the S-CSCF knows the state of each session and registration, it is also capable of updating other entities that may have subscribed to event notification using the SIP *SUBSCRIBE* method. Whenever the state of a subscriber device changes within the network, the S-CSCF notifies all entities involved of the state change.

This allows services such as presence to be notified if the status of a subscriber's registration should change, while also preventing unauthorized access to the registration status. Only authorized entities are allowed to use the *SUBSCRIBE* method, as validated by the S-CSCF. This also eliminates the need for the device itself to send registration status to multiple entities within the network.

In short, the S-CSCF is the nucleus of the IMS. It is the core of the network, providing the one point of control within the network that enables operators to control all service delivery and all sessions. Of course, there is never just one S-CSCF unless the network is very small, so this is a distributed function. The S-CSCF should be deployed in the network in accordance with the number of subscribers to be maintained and the types of services that are to be supported and controlled by the S-CSCF.

Subscription Locator Function (SLF) This function can be built into another function, or it can be a stand-alone server. The purpose of this function is to locate the HSS and S-CSCF assigned to a particular subscriber. This is an indexing function, mapping the user identity to the S-CSCF/HSS according to registration.

When the P-CSCF needs to route a request for a subscriber session to the appropriate S-CSCF, the P-CSCF would access this function to determine which S-CSCF has been assigned to the subscriber. Other devices may need to access this function as well, such as an application server supporting services to the subscriber.

Home Subscriber Server (HSS) While the S-CSCF acts as the core of the network, the HSS (shown in Figure 1.5) serves as the central source of subscriber data. The HSS stores user data such as the services a subscriber is allowed to access, the various identities (the private user identity and all public user identities), the networks the subscriber is allowed to roam to (in the case of wireless), and the location of the subscriber device.

When a subscriber registers with the network, the S-CSCF accesses the HSS to obtain the user profile. The user profile is what identifies all of the private and public identities associated with the subscription, as well as the service profiles for each of the identities.

Whenever there is a change in the subscription of a subscriber device, the information is pushed to the S-CSCF. The HSS will send all of the subscription data to the S-CSCF, not partial data. This eliminates the possibility of data becoming corrupt or out of synch with the HSS. The S-CSCF then replaces all subscription data it has with the data the HSS sends.

The purpose of the registration is to provide a location for a subscriber device. Location does not necessarily mean the exact location. In the case of wireless, the location can be the GPS coordinates, but it usually means the cell site identifier. In wireline networks, the location depends on where the P-CSCF that was used to access the network is located. The location is also the IP address assigned to the subscriber device. In other words, the location identifies how to route sessions to the subscriber at any moment in time.

This is why registration is so dynamic. Each time the subscriber changes locations, the address assigned at the IP level is changed and must be updated within the registration so that sessions can be connected properly. This is analogous with wireless networks where the handset registers with the HLR/ VLR functions.

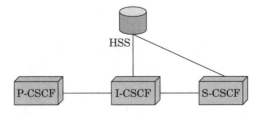

Figure 1.5 The HSS within the IMS

Also a part of the registration information is the subscriber data. The private user identity and the public user identities are all stored as part of the registration data. This allows the S-CSCF to provide full support to the subscriber device and any services it wishes to use.

Services are identified by their service identifiers and are stored by public user identity. As you will see in Chapter 3, public user identities can be assigned to multiple service identifiers, based on the subscription. This information is stored in the HSS, and is shared with the S-CSCF when the subscriber device registers with the network.

If a subscriber is to be barred from service access, the operator is able to bar the public user identity or the private user identity associated with a subscription at the HSS. This provides a central location where the subscription can be controlled. If the subscriber is roaming in another network, the barring of services still applies, since the network they are roaming in will query the HSS via the S-CSCF to determine what services the subscriber should be allowed.

One of the most critical functions of the HSS is to provide the encryption and authentication keys for each subscription. When the subscriber device registers with the network, the assigned S-CSCF challenges the device for the proper credentials stored by the HSS. The S-CSCF queries the HSS to determine what the proper credentials should be during registration. The subscriber device then returns a second *REGISTER* message containing the proper credentials.

Only the network provider and the subscriber device know what these keys are, as they are programmed into the device itself (more specifically, the SIM card of the device) and the HSS. This is another reason for placing the HSS and S-CSCF functions in the core network.

If you are familiar with GSM networks, then the HSS should be a familiar entity. This is relatively the same as the HLR found in today's GSM networks. There are several HSSs deployed within any one IMS, based on the number of subscribers that must be supported within the network. Since this is where subscriber data is located, the HSS must have the capacity to support the subscribers it will be assigned.

However, the HSS is not made available to other networks. Only the S-CSCF within the same network should be allowed to access the HSS within any network. This is because of the subscriber data stored within each HSS. Providing access to a non-trusted domain would be a breach of security and would open up opportunities for subscriber identity theft. The P-CSCF and the I-CSCF guard the S-CSCF and the HSS from unauthorized access, as I have discussed earlier.

Think of the HSS as the brains of the operation. Any service that a subscriber is privileged to use can be found in this one central location. All changes to the subscription are made in one location. This is a stark difference from VoIP implementations where the subscriber and his or her privileges are managed through various softswitches, in a mesh configuration, or using application servers accessible by the subscriber.

Another advantage of the HSS is the ability to manage multiple identities under one common subscription. As discussed in Chapter 3, a subscription may have only one private user identity, but it can have multiple public user identities. Service identifiers can also be assigned to each of the public user identities based on the one subscription. Read Chapter 3 for more details on user identities and how these are used within the IMS.

Application Server (AS)   The Application Server (AS) is a multipurpose function within the IMS; however, it is not part of the IMS core (see Figure 1.6). The AS speaks SIP, as well as DIAMETER, and interfaces directly with the various CSCFs within the network.

There are many uses for the AS, such as applications, service delivery, content delivery, presence server, and even video conferencing. The AS can create a SIP dialog, which means it can behave as a SIP user agent (UA). There are scenarios (depending on the service being provided) where an AS may even communicate with other Application Servers. For example, the AS can act as a redirect server generating requests and sending responses to other SIP entities within the IMS.

The services provided on an AS are identified by service identifiers. Think of these as addresses for the services. These in turn correspond to entries within the HSS where users are registered. Each subscription is defined with all possible user identities and with the service identifiers associated with each of the user identities. This is what allows the S-CSCF to route service requests to the proper AS. The identifier is communicated as part of the *INVITE* request.

If an AS has subscribed to event notification (via the SIP *SUBSCRIBE* method), the S-CSCF will send the *REGISTER* to all of the Application Servers that have subscribed to notification. This allows the servers to stay apprised of the users status at all times. The AS does not generate a *REGISTER* message, however, as it does not need to register with the S-CSCF. While the AS recognizes the *REGISTER* method, it does not process the message other than by recording the status; the S-CSCF is the SIP registrar within the IMS domain.

Media Gateway Control Function (MGCF)   This function is mentioned here only because it does provide access into the IP/SIP domain, although it is not really considered part of the IMS (see Figure 1.7). The MGCF connects into the Public Switched Telephone Network (PSTN) domain, providing a gateway function between the SS7 world and the IP/SIP world.

When a call originates in the SS7 side and terminates in the SIP domain, the SS7 messaging (ISUP) is sent to the MGCF, which is responsible for mapping the SS7 message (or BICC) into a SIP request prior to forwarding on to the P-CSCF.

Remember that the P-CSCF does not support SS7; it is only a SIP device, therefore there must be a function that provides the conversion between networks. This is one means of providing this, while at the same time controlling the bearer path.

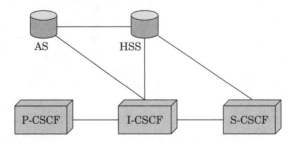

Figure 1.6   The Application Server (AS) within the IMS

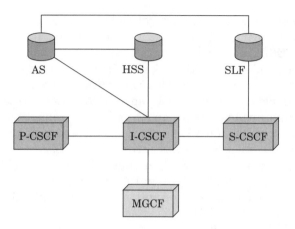

Figure 1.7 Location of the MGCF

The bearer path consists of a trunk that terminates into a media gateway, under the control of the MGCF. The trunking is controlled per the ISUP messaging and passed via a media gateway to the IP network under the control of SIP. This would be common in networks with well-established VoIP networks.

Another way of providing conversion between the PSTN and IMS domains is through the use of a signaling gateway. This is useful when there is no VoIP network already established. The trunking then is managed by the network switches and eventually must be sent to a packet network, although the packet network could be provided by another service provider.

The primary purpose of the signaling gateway is to support SS7 over IP networks. SS7 is a packet protocol, but the transport and network layers of the SS7 protocol (the message transfer part) were designed for TDM facilities. The Internet Engineering Task Force (IETF) has developed new transport and network layers for SS7 ISUP and SCCP/TCAP using the SIGTRAN protocol on IP facilities.

The signaling gateway then receives the SS7 traffic over TDM facilities, next converting the MTP layers to SIGTRAN for transport over an IP network. The payload of the SIGTRAN packet remains the SS7 ISUP and SCCP/TCAP protocols, so no conversion is made from SS7 to SIP at this function. That remains the responsibility of the MGCF.

The MGCF does not replace the P-CSCF; rather, they work in conjunction with one another as seen in Figure 1.7. The MGCF provides the connection to the SS7 network, while the P-CSCF then takes over the SIP dialog as the session is forwarded into the IMS domain. The MGCF only interfaces to the SS7 network within the same service provider domain. In other words, if a service provider has both an IMS and a legacy network, the MGCF interfaces between the two network domains, but not to other service provider networks.

Breakout Gateway Control Function (BGCF) The BGCF works like the MGCF but is a gateway to other providers' networks (see Figure 1.8). The MGCF interfaces to the

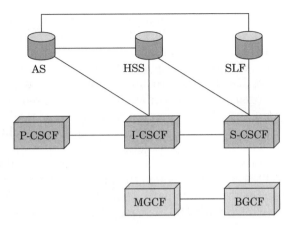

Figure 1.8   Interfacing networks with the BGCF

SS7 network within its own domain, but when it needs to connect to another service provider network, it interfaces through the BGCF.

The BGCF provides a gateway function between networks. For security reasons, it should only interface with other BGCFs in other networks. The BGCF will interface directly with the S-CSCF within its own network, so that when a request or response is to be forwarded to another network, the S-CSCF can route the message directly to the BGCF for routing to the other network.

The S-CSCF routes to the BGCF according to the TEL URI within the request or response. Routing tables within the S-CSCF identify the proper BGCF according to TEL URI. When routing a session into the PSTN, the S-CSCF has no knowledge of routing within the PSTN. It relies on the service of the BGCF to route into the PSTN.

Media Resource Function (MRF)   There are two functions within the MRF; the media resource function controller (MRFC) and the media resource function processor (MRFP). The processor manages the various media types under the control of the controller.

Examples of media controlled by the MRF are as follows:

- Tone generation
- Conferencing
- Text-to-speech
- Tone detection
- Automated speech recognition (ASR)
- Facsimile
- Connection control
- Announcements

The MRFC controls media streams in the processor by interpreting requests sent by an application server or an S-CSCF. The media themselves may be mixed within the MRFP, for example as in a conference call. The MRFP also controls flooring in a conference session. The AS provides start/end times for the conference, participant addresses, and possibly host data to be shared during the conference session.

A Webinar is a good example. Different media are used during a Webinar. Voice, data, video, and control are all incorporated into the same session. The AS provides the addresses of all the participants and will even host the Webinar application. The MRF controls the various media as just defined.

If providing tones, announcements, or conference capabilities, the S-CSCF or the AS sends an INVITE to the MRF, which then responds and delivers the tone as requested. The SDP identifies the bearer path where the tone is being delivered. The MRFC responds to the S-CSCF with 100 Trying. The MRFC acts as a terminating user agent in these scenarios. Conference identifiers are sent in a 200 OK response.

When granting control of the floor during a conference session, the MRFC sends a "re-*INVITE*" to the user being passed or granted control. Flooring is the function used in conferences that allows another participant to control the application. This is commonly used when showing presentations or sharing documents, for example. All of the other participants then have the ability to see what is being shared, but they cannot control the session until the host grants them the floor.

IMS Reference Points

The protocol used throughout the IMS is SIP; however, there are instances when DIAMETER is used. Any instance when session control is needed, SIP is used. When accessing subscriber data or charging data, DIAMETER is used. This would mean that some devices must be able to communicate using both SIP and DIAMETER.

Figure 1.9 depicts the IMS reference points for both wireline and wireless. These are defined by the 3GPP and represent the major interfaces. Additional views of these and other reference points are provided here for clarity and to provide other views of how IMS may be implemented (in an all-wireless network, for example).

Figure 1.10 depicts an all-wireless implementation. This also depicts the evolution of GSM wireless as defined by 3GPP. When looking at the GSM network, one can see where GPRS was added for data support. GPRS provides a network where data can be offloaded from the wireless switches, and also provides some additional bandwidth at the radio interface.

The GSM radio reference points are identified as single-letter identifiers. The predominant protocol used over these interfaces is the Mobile Application Part (MAP). The E interface sits between two MSCs within the same network. The Gateway MSC in this figure identifies a gateway function interfacing into another GSM network.

The MSC in turn interfaces with the *base station controllers (BSCs)* in the network via the A interface. The A interface allows the mobile switches to communicate with the radio resources in the network using SS7 signaling, as well as bearer path (voice channels from the radio resources into the mobile switching center).

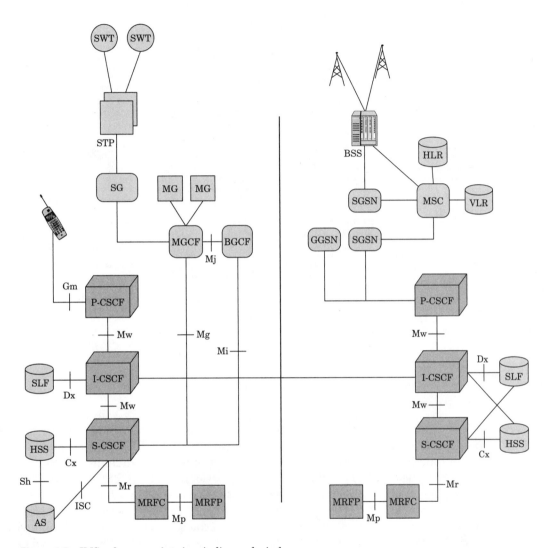

Figure 1.9 IMS reference points in wireline and wireless

The BSCs interface to *base transceiver stations (BTSs),* which provide the GSM radio function. The BTSs provide the radio functions interfacing to the antenna(s) at the cell site, under the specific control of the BSC. They communicate to the BSC via the A-bis interface.

For accessing subscriber data, the MSC must communicate with the HLR. This is supported via the D interface. The VLR function is typically integrated with the MSC, even though here it is depicted as a separate entity. The H interface is used for communicating with the Short Message Service Center (SMSC), used for managing short messaging services in the network.

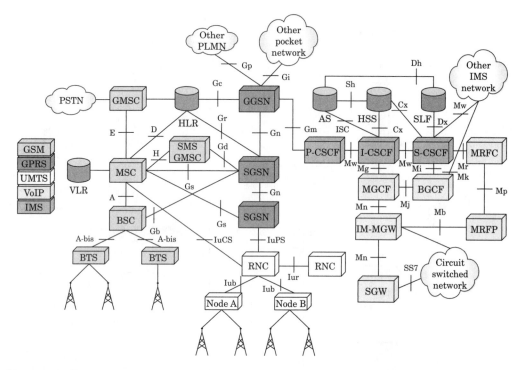

Figure 1.10 Wireless reference points GSM to IMS

The GPRS interfaces are identified by a *G* followed by a lowercase letter. The GPRS protocol is used over these interfaces. Basically, the GPRS overlay adds two additional functions to the network for the purpose of packet switching. The Serving GPRS Support Node (SGSN) provides packet connections from the BSC via the Gs interface, connecting to other SGSNs in the same network via the Gn interface.

When connecting to packet services outside of the network (such as another GPRS network, or the public Internet), the SGSN interfaces to a Gateway GPRS Support Node (GGSN) via the Gp interface. The GGSN interfaces to other wireless GPRS networks via the Gp interface. The Gi interface provides access into another packet network. The Gm interface provides access to a P-CSCF in the IMS domain.

The UMTS interfaces identify additional radio bandwidth for supporting broadband cellular services. All UMTS reference points begin with the letter *I* followed by a lowercase *u* and one or two additional letters. For example, the IuCS interface provides connectivity from the Radio Network Control (RNC) to the circuit-switched network (through the MSC). The IuPS, on the other hand, provides an interface to the packet-switched network.

Therefore, voice traffic is routed from the UMTS network via the IuCS interface while data is routed through the IuPS interface (through the SGSN). The Iub interfaces to the radio nodes, while the Iur interfaces to other RNCs in the network.

So far I have just described 3G wireless. The IMS is the next phase of wireless, depicted to the right of Figure 1.10. Notice that the figure also includes *Media Gateway Control Functions (MGCFs)* representing VoIP connections, even though we are talking about wireless evolution. These would interface to any ISUP (or BICC) trunking supported in the network. The IMS interfaces are prefixed by M followed by a lowercase letter. SIP is the protocol used over these interfaces.

The DIAMETER protocol is used over interfaces Sh, ISC, Dh, Cx, and Dx. Diameter can also be found in the charging architecture as described in Chapter 6. Remember that DIAMETER does not carry session control. It is used for sharing subscriber and charging data in the IMS domain.

The reference points defined in this chapter provide an outline of all of the major reference points within the IMS domain. There are others defined by 3GPP2, which is focusing on the CDMA implementation of the IMS. The model is very similar to the model and reference points described in this chapter, with additional functions specific to CDMA networks.

As one can see throughout this chapter, the IMS represents a new architecture for all networks, but not a network so different than what was implemented to support the IN and SS7. There are many very similar concepts.

Also remember that this architecture was developed as an evolution of the GSM network architecture. One of the drivers of redefining networks was to ensure that security and reliability were preserved when implementing IP in the wireless domain. Conventional VoIP implementations were inadequate in providing security and did not interoperate well between various vendor nodes.

This led to many different problems with VoIP implementations within wireless networks that were already complex. Since authentication and registration were already concepts implemented in GSM, the 3GPP community felt these functions were essential in any IP implementations, and therefore set about to provide the architecture to support these functions at a minimum.

The IMS provides both wireline and wireless operators with a path to evolve their current networks to an all-IP services network, while enhancing their current security. Certainly many wireline networks would benefit from authentication as well as the call session control functions defined by IMS.

Moving from Legacy to Convergence

As should be expected, migrating a network from a legacy architecture to a next-generation technology is not a small undertaking. It involves so much more than simply launching a few platforms, connecting up the circuits, and calling it a network.

There are countless systems and applications that are all connected in some fashion to the network as well as the various network elements. These systems must somehow interoperate with all the other systems and applications, to ensure a seamless transition. Throughout the entire process of transitioning the network, the operator must somehow ensure that subscribers never experience a degradation in service, and that all investments are protected and future-proof.

When the telephone companies of the world implemented Signaling System #7 (SS7) and the Intelligent Network (IN), their intentions were certainly noble. Yet SS7 and the IN concept failed on a number of fronts in my mind. There was so much emphasis on the "killer application" that would somehow justify the expense of implementing such an infrastructure, that the entire industry simply set itself up for disappointment.

The concept itself was a great idea and is certainly still worth pursuing in today's modern networks. You will see constant themes of the IN throughout this book as we discuss the various functions and features of the IMS, but the implementation needs to change.

For example, the IN called for services and applications to be located in the core of the network instead of on the network switches themselves. The idea was to reduce the cost of supporting services and applications by eliminating the need to replicate the service at every single network node.

However, for this to work, special software was required at every switch responsible for delivering the service. The switch software was responsible for accessing the service control point (SCP) responsible for providing the service. The software "triggers" became extremely expensive, and therefore it became prohibitive to launch some services on a grand scale in the smallest of networks.

When the public began pushing for more mobility, the technologists of the industry once again demonstrated how this could be accomplished using the existing architecture of the world's networks, utilizing SS7 and the IN. But the cost to implement such a service was just too prohibitive and therefore prevented the world from realizing this feature.

VoIP enabled mobility, demonstrating that such a service was possible using the Internet technology and was a viable service offering for even the smallest of service providers. However, the issues of security were troublesome for traditional operators, making VoIP a troublesome alternative.

The industry also learned that depending on a vendor's end-to-end platform solution was costly. SCPs, for example, would only run the software provided by the vendor selling the hardware. An SCP could cost millions for a simple service like Calling Name.

The industry is attacking these problems aggressively, as it works at redefining the world's communication networks, as well as the business model that funds them. Application platforms, for example, are being standardized on generic platforms available from any vendor at a fraction of the cost of a traditional SCP. By standardizing these platforms, operators will be able to purchase an application server from any vendor they choose, and place any vendors' software on that platform.

This means operators are no longer married to a single vendor for their service platforms. They can pick and choose from hardware vendors and software solutions to meet their business needs. This opens up a whole new market for software vendors who could not previously compete against the platform giants.

While it has been proven that voice services can be offered over the Internet, there is still work to be done to make this as robust as the traditional networks we use today. Of course in the end it may not really matter to the key buyers of communications services. The younger generation does not accord service reliability as high a priority as their parents, as demonstrated by their choice not to own a landline at home. They are perfectly content using their cell phone for all their needs, even if the service quality is not the same as what they could get from landline service.

The transition to an all-IP network means making some changes to the very technology that makes the Internet possible. These changes will make the technology all that much more robust and "carrier grade." But more important, the implementation of this technology is being well defined, providing operators with a clear and reliable means of implementing IP throughout the network, delivering Internet-like applications to all subscribers, while still maintaining the very business model that funds the network to begin with.

This chapter addresses the migration from a legacy network to an all-IP network, using planned phases rather than "flash-cutting" the entire network. The discussions throughout this chapter are based on what the author has witnessed from the world's leading carriers.

Migrating the Backbone to an IP-Based Network

There are stark differences between wireline and wireless in terms of migration plans to an all IP–based network. The wireline operators of the world began in the 1960s with plans to convert the backbone of their networks to packet-based rather than circuit-switched, to address the growing issue of maintaining many different levels of network for different traffic types.

Their migration has been more chaotic than wireless, often choosing new technologies as they came along, rather than crafting a long-term migration and implementation plan with a common goal in the end. Many times it appears as if the wireline operators are simply choosing the technology of the day to meet their packet requirements.

The wireless operators, on the other hand, have an orchestrated, laid-out plan and strategy for moving the world's wireless networks to packet. Consider the work of the 3GPP and the evolution of the GSM specifications. In reality, both are probably about the same when it comes to choosing the latest technology, but the 3GPP has certainly chosen an orderly migration path.

The migration from TDM-based networks to packet-based networks began some time ago. Bell Laboratories began development of Asynchronous Transport Mode (ATM) as a new backbone technology, enabling the Bell System Companies in the U.S. to move all of their bearer traffic to a packet-based network rather than a circuit-switched one.

ATM eventually moved out of the labs and began the long road of implementation in the world's networks. While largely successful, many may argue that it is still cost-prohibitive when compared to the cost of IP. Still, IP cannot offer the same reliable service of an ATM transport network, a fact that has slowed the implementation of IP in many traditional telephone networks.

The primary issue is the lack of support in TCP/IP for real-time traffic such as voice and video. TCP/IP was developed simply for the transport of data and is far more tolerant of delays and packet loss. Voice and video, on the other hand, are not so tolerant, which is why ATM was adopted so quickly.

The ATM protocol answered the primary need for all operators: Quality of Service (QoS). By laying ATM on top of a SONET fiber backbone, operators can realize a very robust packet-based backbone capable of moving all of their aggregate traffic.

However, ATM is not the best protocol for voice and video for many reasons. Developers realized that if they developed a protocol specifically for voice and data, it would consist of smaller packets delivered much quicker to the end points. This theory does not work, however, for data networks.

When data is sent using smaller packets, it takes many more packets for the successful transfer of large data files. This is not only inefficient; it can create a lot of overhead within the network. ATM was a compromise between the two; providing smaller packets for voice and video, eliminating the delays in packet transmission, while offering large enough packets to support the transfer of larger data files.

TCP/IP development has continued as well. The Internet Engineering Task Force (IETF) realized that the TCP protocol is not a good transport for real-time applications. They set off to create a new peer protocol referred to as Stream Control Transport Protocol (SCTP). This protocol was first used in support of SIGTRAN (SS7 transport over IP) and has since expanded to other uses within the VoIP domains.

Many of the shortcomings of IP have been addressed, and service providers have quickly learned the economies brought about by an IP infrastructure. Yet providing highly reliable and secure telecommunications services over an IP network requires some changes to the technology as well as the implementation of IP networks.

One of the key issues with VoIP is really an implementation problem. There is lacking a set of implementation/interoperability standards for the deployment of a VoIP network.

For example, an operator may choose to deploy a media gateway controller (MGC) for call and session control, but it has many choices for the session control protocol.

Even if an operator selects SIP, there is little defined in the way of implementation for SIP today to support security. Certainly if we investigate the breaches that have been committed today in VoIP networks, we will find that these breaches were a direct result of implementation weaknesses.

Quality of Service (QoS) and bandwidth management are a couple of key areas that have required work in the past as well. Operators have adopted numerous methodologies for dealing with these issues and have begun transitioning their core network to IP.

Before an operator can take advantage of IP services, it must first convert the backbone network to packet-based. Look at how the GSM network has evolved. Wireless provides a good model for this discussion because of the planned and methodical approach taken by GSM operators to evolve the networks to support packet-based services.

GSM operators first added an overlay network to their existing infrastructure for data traffic. The base station controllers (BSCs) moved packet data traffic away from the mobile switching centers (MSCs) to this packet data network (the General Packet Radio Service, or GPRS).

The GPRS network then provided IP facilities into the Internet and other IP-based networks. This allowed the operators to make investments in their packet networks without having to expand the capabilities of the voice switches in the network.

Wireline operators have followed this path as well, deploying IP in their backbone transport networks, and moving bearer traffic over these transport networks. This is the first step toward an all-IP network.

This is where the VoIP elements come into play. In order to support the IP backbone, there needs to be conversion of traditional voice to packetized voice. This is the job performed by media gateways, under the control of media gateway controllers.

VoIP networks were developed to convert voice transmission from its traditional analog format into a digital packet format that could be transported over the new IP backbone. However, there is more to this than simply converting the voice to packet formats. Voice transmission is not tolerant of delays. If there is a delay in the delivery of the voice packets, conversation becomes intolerable.

This is the challenge faced by many VoIP providers. While many have solved the latency and QoS problems for peer-to-peer calls, there is still work to be done for calls that begin in one network and are transported to IP networks.

VoIP introduced many other challenges as well, but this is outside the scope of this book, so I won't go into those details. I will mention here though that these challenges are partly what drove the 3GPP to begin development on IMS.

But more than just voice services rely on the backbone network. Signaling is also transported through the backbone along with the voice traffic. Since signaling (SS7 specifically) is already packet data, there is no need to convert the signaling data to packet. However, the SS7 protocol relies on transport protocols that were developed for use on TDM facilities. The transport layers of SS7 (the Message Transfer Part) will not support IP facilities and therefore must be replaced with a new transport protocol before SS7 can be transported via IP.

This was the work of the IETF, which developed the SIGnaling TRANsport (SIGTRAN) protocol. SIGTRAN replaces the MTP layers of the SS7 protocol with transport protocols developed specifically for IP networks.

These protocols emulate the MTP services, over IP networks. This is necessary for a number of reasons, QoS within the signaling networks being one of the primary drivers. There are many procedures provided by level 2 and level 3 of the SS7 protocols to ensure the availability of signaling links, as well as the integrity of the signaling data itself. One of the simplest of these procedures is the transmission of a fill-in signal unit (FISU).

The FISU is really just a protocol flag carrying no real information. However, its absence signifies trouble with the transport (the physical link). The main purpose of providing this as a function is to provide a proactive, rather than reactive, detection mechanism for link failures.

In other words, instead of finding out that a link is out of service when there is data to send, the protocol and all of the network entities use the FISU as a link integrity measurement. As long as the entity is able to receive and process the FISU, the link is operating correctly. If the entity is no longer receiving FISUs, there is something wrong with the link.

The links themselves are always transmitting something. If there is no signaling data to be delivered, then FISUs are transmitted over the idle links. The entities are always checking the health of the link even when there is no data being sent.

This proactive approach to maintaining transport reliability is seen at level 3 of the SS7 protocol as well. When a link fails, there are automatic procedures to ensure the orderly routing of signaling data around the failed routes or links. This improves the availability of signaling routes by ensuring packets are rerouted through good routes when other routes and links fail.

All of this is automatic, of course, and many times when a link fails, it is brought back into service before anyone even has time to react. These processes have to be emulated in an IP environment, to continue to ensure the availability and reliability of the signaling network.

There are several protocols provided in the SIGTRAN protocol suite:

- M2UA
- M2PA
- M3UA
- SUA

SIGTRAN is used to support the transport of SS7 level 4 signaling data (ISUP and TCAP) over an IP facility. The payload of these SIGTRAN protocols is still ISUP and TCAP, so the entities themselves must still support SS7. There are also specific entities that interface with the IMS that support SIGTRAN and provide the conversion between the two signaling types, as we will discuss in later chapters.

One of these functions is the signaling gateway (SG). The signaling gateway is responsible for interfacing into the SS7 network, and converting the MTP layers to SIGTRAN back into the packet network. The ISUP and SCCP/TCAP layers remain intact and unaffected by the changes.

The signaling gateway is another means of evolving the network from a TDM-based core to an IP-based core. For example, an operator may decide to maintain its existing TDM facilities in the core network, while at the same time implementing an IP network as an overlay. As new entities are added to the network, they are added on the IP side of the network, using a "cap and grow" approach rather than a replacement strategy.

Using the signaling gateway at the edge of the TDM networks provides a bridge into the IP network. The operator can then depend on media gateway controllers (MGCs) to accept the SS7 signaling, and eventually make the conversion to SIP.

The conversion to SIGTRAN is an important stepping stone to full network transition, since signaling performs the important function of call control. Converting the signaling facilities to IP provides many economies to the operator and allows it to quickly realize additional capacity and throughput not present in its legacy signaling networks. The business case to IP backbone, therefore, becomes much easier to justify when built around cost savings, and network growth with added density.

The signaling gateway only converts the transport layers to SIGTRAN/IP. Since ISUP and SCCP/TCAP are not supported in IMS, these layers must also be converted to the Session Initiation Protocol (SIP). This conversion is provided by the media gateway controller, which accepts the SIGTRAN/SS7 signaling from the legacy network and converts this signaling to SIP-based signaling toward the IMS. This is one of the reasons operators should contemplate VoIP deployments as the next phase toward IP-based IMS.

A VoIP deployment will consist of all IP entities supporting packetized voice and data across an all-IP backbone. However, the SS7 legacy must still be supported even in the VoIP domain. Calls originating in the TDM domain will cross boundaries, as will the SS7 signaling associated with these calls, requiring support of SS7 even in the IP domain for some time to come.

Deploying VoIP as a Growth Strategy

Besides the obvious that Voice over IP (VoIP) provides an economical means of growing the network, the VoIP functions (media gateway and media gateway controller, specifically) provide the next phase of implementation needed to support a full IMS network.

Consider this: the TDM network relies on switches, which provide all of the functions needed to originate and terminate a telephone call. These switches are expensive today and are difficult to cost-justify in small rural markets.

However, if an operator could deploy a small, inexpensive box consisting of nothing more than a switching fabric (the matrix connecting one circuit to another), the operator should be able to reduce the cost of their service offering in those markets.

Removing the "intelligence" from the switch means placing the call control in the core of the network. This was the original concept behind VoIP networks. Place the switching fabric out at the network edge, but keep all of the intelligence and call control in the core where it can be secured and maintained.

The MGC is the entity bringing call control to the core of the network. This is the entity that interconnects all of the media gateways together, and using a call control protocol (such as SIP) communicates to each of the media gateways what resources (such as codecs) are needed to support a call or session.

The voice traffic still needs to be packetized, and network growth needs to be supported without significant investments in legacy network equipment. The purchase of additional legacy switching nodes does not make sense if there is a long-term plan to convert to IP, and VoIP can support the voice network as the operator begins its migration to IMS.

The transition at this first step is usually at the tandem level (back in the core of the network). Then as subscribers are added, or legacy equipment needs replacement, those replacements are achieved using VoIP deployments (MG/MGC). This now takes care of packetizing the voice portion of the network and supports slower markets even after IMS is fully deployed. The operator can then slowly begin retiring those legacy switches and replace them with media gateways, while extending the reach of its IP facilities to all of their markets.

Of course, we would be foolish to think that voice is the only reason to migrate to an IP infrastructure. Data is a natural fit for the IP backbone (since this is what IP was developed for), but video can be delivered over the IP network as well.

At this point, the operator has implemented IP in the core backbone, migrated its signaling to SIGTRAN/SS7, and begun deploying VoIP at the core, and then the edge of the networks to support packetized voice. Now all that is left are the service platforms and the operator is ready to declare itself a convergent network, right?

Wrong. The missing link here is still call (or session) control. While it is possible to implement an all-IP network and support any service an operator wants today, managing these networks is an absolute nightmare.

Look at companies such as Skype and Vonage. They began using proprietary architectures and session control, and quickly found out that these approaches did not scale very well. They are now in trouble as they struggle to rebuild their networks in response to tremendous growth they did not anticipate.

Add to the picture billing and partner management, and you can quickly have an absolute nightmare on your hands. Let's looks at what is happening with wireless today. I have spoken with numerous wireless carriers who are providing music downloads to their subscribers. The service delivery is the least of their problems.

They found that their existing billing systems did not support this type of service, and so they had to build a completely new billing mechanism to support music downloads. This new billing system had to somehow be integrated with their existing billing systems (many operators have shared with me that they have as many as 15 different billing systems, all supporting different disparate services).

So while it is true that an operator can deliver any service it needs using an IP network, it faces many challenges when it comes to managing that network, and maintaining a secure, reliable network. This is where the IMS comes in.

Deploying IMS

The business case for deploying IMS is not about the services it provides (there are none), nor is it the killer applications (there are none). The business case for implementing IMS is a little more difficult than that, but all the more rewarding long-term.

Implementing IMS is about supporting a single architecture for all services, rather than multiple systems for multiple services.

For example, let's say you are an operator with a traditional IN-based network (using SS7 call control/signaling). You now want to offer your subscribers a new service for music downloads. Since SS7 does not support this type of service, you will have to find an alternative (most likely an IP-based service delivery platform with this capability).

Your billing system is the next hurdle. Your present billing system supports CIBER/TAP3 records from your switches, but it does not support IP-based services, so you will have to add a new billing system as an overlay to your existing system. This will somehow have to be integrated into your present billing presentment system so that the proper charges can be applied to the subscriber bill.

To make matters worse, your present billing mechanism (and all audit functions of that billing system) are based on a minutes-of-use model. Music downloads are not billed this way. Subscribers pay for the music download one time, for a flat fee. In some cases, you may want to issue a license where the subscriber pays for a period of time. At the end of that period of time, the subscriber has to pay another cycle.

Your legacy systems do not support this model, and hence the need to add new systems that do not integrate very well with the legacy systems. You end up managing these separately.

You now have the billing taken care of, but you have no means of auditing the transactions between the subscriber and your content delivery platform. Worse, the content is owned by another content provider, so you must interface to its server to purchase the content, and then deliver the content to your subscriber. The content provider must be paid at the time of your transaction, but the subscriber is not going to pay before receiving a bill from you at the end of the month.

When the subscriber finally receives their bill, that subscriber refuses to pay, because he or she never received the content. You do not have the tools to see the transaction when it took place because your existing monitoring system does not support IP networks and does not provide visibility to the IP transactions with the content provider. You have to purchase yet another system to monitor the IP portion of your network.

And of course, we haven't talked about the network requirements to provide the transport of music to your subscriber. Delivering music downloads over IP is simple, yet it can be complicated if you are trying to ensure that the music is indeed received by the subscriber. If you are relying on the Internet model for this service delivery, be prepared to lose revenues from piracy (already prolific over the Internet). Now you have to add a digital rights management system (DRM). Which of course cannot be integrated with any other systems in your legacy network.

This is not so far-fetched. There are many operators today delivering multimedia services using their traditional networks with IP overlays. However, when one tries to support all of the back office requirements and other support systems needed, it quickly becomes a nightmare to support.

This is why the IMS is a better alternative than traditional approaches, or even pure IP-based networks. The Internet model works great as long as you don't care about things like security, business intelligence, customer care, fraud, and billing. When it comes to these things, the Internet model becomes much more difficult and costly to implement.

The IMS, on the other hand, is designed to support all media, and all services, using one common architecture and one common signaling method. This simplifies all of the other functions needed by an operator to support its network and its services.

It also reduces the overall cost of supporting multimedia. A service delivery platform can be reused for many different services, all running through the same facilities. All of your platforms are interconnected through the same infrastructure, making it possible to offer unique, value-adding services such as presence.

The ability to begin integrating various services is an important factor to consider when making a decision toward implementing IMS. Tying many different platforms together to support a single service is no small feat if all of these platforms rely on proprietary protocols. IMS supports this capability by normalizing all session control through SIP. This is perhaps the killer app for IMS.

Changing to IMS does not happen overnight. There are many legacy systems that will have to interoperate with the IMS network as it is implemented. Some operators have chosen to treat their IMS deployments as green-field networks, running them completely separate from their legacy networks. This can work, but it leaves a lot of legacy equipment stranded with no opportunity of recuperating the capital investments.

Others have chosen a more transitional approach, using the legacy network and the IMS together, with gateways and other elements in between the two networks supporting the interworking between the services layer and the transport layers. This approach ensures subscribers continue enjoying the services they have now, while providing them the opportunity to enjoy new multimedia services, without abandoning any part of the network.

This certainly makes the best business sense. Simply "turning away" from the legacy network and dumping all your investment into an IP infrastructure is not smart business; it quickly strands your core subscribers on the legacy side with non-competitive services. Operators need the ability to continue offering their legacy customers new, innovative, and competitive services through strategic investments in their legacy networks.

At the same time, new services should be offered on IMS-ready platforms to prevent the stranding of those investments. This ensures the platforms the operator invests in will make long-lasting investments supporting an all-IMS environment when the time comes.

Most will agree that if a legacy network exists, the best approach is to transition from an IN-based network to IMS. This is much easier to accomplish in the long run when using solutions available today to bridge between the old and the new. The sections that follow provide more detailed information about how these networks interoperate with one another.

Interworking Between the Legacy Network and IMS

Within the core network there are two parts of the network that must be supported. The transport of bearer traffic is the easiest. Once bearer traffic is packetized, moving it through an IP infrastructure is much simpler (yes, I know I am oversimplifying things here) than maintaining a TDM circuit-switched network.

The voice is in a form that must be converted to packet before it can be sent through the IP infrastructure. The IMS itself is not really a part of this, since the IMS is really the call control portion of the network. The IMS itself is the architecture that allows the various packet elements within the network to communicate all aspects of the sessions and share this information between multiple entities.

The second challenge comes in the form of signaling, which is really session control. Session control ensures that subscribers are able to access the same services they enjoy in their home networks anywhere they connect. It is what ensures a subscriber is legitimate and is authenticated prior to connecting to your network. Session control is about maintaining complete control over every service you deliver in your networks, enabling security, billing, and much more. Signaling and session control are how network entities exchange information about a session in an effort to be able to support the session.

This means converting the signaling plane from SS7 to SIP after implementing IP as the transport for the signaling network. SIP is the call control signaling protocol for the IMS. It replaces SS7 in the control plane, supporting all forms of media in addition to voice. This model is very similar to the Intelligent Network (IN), where SS7 is used to communicate between network entities for the support of voice.

The IMS provides the same functions theoretically as the IN used today, except the IMS supports more than just the voice services. It supports all media types and all services within the network, using the same signaling and the same call control.

Again, it is not foreseen that the service providers of the world are simply going to abandon their SS7 networks and start replacing them with SIP. This has never been the case historically as new technologies came along. Certainly SS7 and the IN were not implemented overnight. Operators took many years to implement their SS7 networks, implementing small portions of the network at a time, until eventually the entire switching network was interconnected with SS7 signaling.

So far we have only discussed the transition of the voice transport and signaling parts of the network. There is much more to a network than signaling and switching. The support systems that are used to manage these networks is perhaps the most important aspect of the business, because it is these support systems that allow an operator to maintain profitability, secure the network from attack and unauthorized access, and accurately bill for its services.

Migrating the OSS/BSS

The jury is still out when it comes to OSS/BSS systems for the IMS. Much is yet to be done to provide a complete end-to-end solution for the various functions within the back office and support systems. Provisioning of the various entities, managing the facilities through inventory systems, and service order provisioning where new services are activated for subscribers are some of the functions required to support any telephone network.

There are some solutions available on the market today, supporting SIP and DIAMETER protocols and providing various functions such as monitoring and provisioning. Many more are being developed.

The charging systems used to support billing are evolving, as are the standards themselves. The 3GPP is continuing development of the charging standards to be used in IMS. Charging must be changed, because the IMS supports all forms of media in addition to voice. Charging systems today are built for voice services, or for event-type charging (such as messaging delivery), but do not integrate very well. The 3GPP has addressed this through a completely new architecture to support the charging for multimedia services throughout the network.

Performance management systems are continuing to evolve as well, but most have not yet been developed to capture end-to-end network transactions. This is a critical factor in maintaining an IMS network; not just for performance management but also for revenue assurance and security.

For example, if you are providing music downloads, provided through a third-party partner, your monitoring system should be able to capture the entire transaction. This includes the signaling from the handset requesting the download (the HTTP session accessing the Web portal and selecting the link to download the music file). The File Transfer Protocol (FTP) session used to download the music file from the content provider's server through your network to the handset should be captured to ensure successful delivery. The entire transaction should be captured on the same system so that you don't have to deploy multiple systems to trace each part of the transaction.

It is simply not enough to see one portion of the network. As an operator, you must be able to monitor everything that occurs in the network; from the handset to the service delivery platforms, back to the handset. This is especially important as the revenue chain shifts from minutes-of-use to content purchases.

There are many different functions in OSS/BSS. I will cover the basic functions, but there are most likely many more within your own networks that need to be updated to support SIP. The basic functions are as follows:

- Network monitoring
- Billing and revenue assurance (collection and mediation)
- Service management

Network monitoring has traditionally entailed connecting probes to signaling links within the SS7 network and collecting the SS7 signaling messages for analysis. These systems typically operate in a real-time mode, providing limited storage (anywhere from three minutes to three days). This is typically fine for most troubleshooting, but some applications require much more storage capacity.

The purpose of the network monitoring system is to report on the health of the signaling network and all of its facilities. This involves alarming when signaling links fail or reach congestion, and reporting on the status of the signaling entities themselves. Over the last several years, these same systems have begun providing *call detail records (CDRs)* based on the SS7 signaling.

These CDRs are then fed to other systems such as fraud management applications and billing systems used for inter-carrier billing. These systems were not designed for the tasks many operators need today, simply because there were no requirements to monitor anything more than the voice setup.

In the IMS, monitoring is much different. The IMS supports all sorts of transactions, and it supports many different types of protocols on the data side (such as HTTP and FTP). SIP is used to provide session control, but there are many Internet protocols that are used to access various files and services that must be monitored along with the SIP.

Also consider that the IMS is all-IP, so tapping into the IMS requires tapping into the routers used to interconnect the various IP facilities. All entities within the IMS are interconnected using routers and IP interfaces, so at the transport layer IP must be the first source for capturing session data.

Obviously, since the facilities operate somewhat differently than traditional TDM-based facilities, the metrics used to measure the quality of the facility will change, but the most significant change is the capacity required. The amount of signaling that takes place in an IMS environment is many times higher than found in traditional SS7 networks. In fact in many trials the network vendor Tekelec has participated in, there has been a ten-fold increase in the signaling volume (just voice sessions alone).

There are many reasons for this such as registration and security procedures. When you add in video, messaging, audio, and any other media session supported in IMS, the volume grows exponentially. This is because the IMS requires much more interaction between the user device and the network. These transactions must be captured by the monitoring system in order for the monitoring system to be able to provide the status of the device, QoS of the network transactions, and the overall health of the entities within the IMS.

One benefit of monitoring IMS networks is that everything is managed in one network architecture with the same signaling protocol. SS7 is strictly voice, but SIP controls everything within the IMS. This can be a significant benefit to companies looking for more visibility into their networks and subscribers' activities.

This provides much more opportunity for monitoring systems. Instead of focusing on the health of the signaling network, these systems need to evolve to support many different functions. The newest trend is to use signaling data combined with subscriber data to create business intelligence. Monitoring systems can and should play a very central role in providing the necessary data and software applications for business intelligence within IMS networks.

This is especially true in IMS because literally everything a subscriber does to communicate is controlled through SIP and the IMS core. This provides an excellent opportunity for the systems collecting SIP signaling to provide this information for business intelligence as well.

Another area in which monitoring systems can play a vital role is in the collection of billing (or charging) transactions. There are two areas where billing information can be found. The first is within the SIP signaling itself. SIP signaling carries several headers used to communicate billing information to various entities within the network.

The billing (or charging) information carried within the SIP headers is then used by the various entities within the network to generate an actual charging detail record.

The charging data does not necessarily provide the information necessary to derive charging and rating, but it identifies the services being charged for and the charging entities within the IMS charging architecture that are responsible for the collection and correlation of the charging data. This information is already being captured by monitoring systems.

The charging entities themselves use the DIAMETER protocol to communicate actual charging data based on what the charging entity derived from the SIP signaling. This means that if the monitoring system is already collecting the SIP information and can support DIAMETER as well, then all that is needed is a charging verification application to compare the two sources for discrepancies.

Auditing of transactions to ensure they were successful (service assurance) and verifying charges are accurate are two very important functions when deploying IMS. They are missing from many traditional networks today, and operators are finding out very quickly as they begin looking into these areas that they are losing much more revenue then they first believed, simply because they have no visibility to the operations within the network.

Operators simply cannot afford to lose revenues because of things they cannot see. Likewise, they cannot afford to lose revenues because of service delivery failures. This is yet another area monitoring systems must be able to support. Since they are capturing the SIP data, all that is needed is an application that provides reporting of service-specific transactions. For example, was a text message delivered successfully or did it fail.

These areas require monitoring of much more than just the IMS domain. The access domain must also be monitored to be able to ascertain where trouble began, and what caused failures. This means supporting GPRS, UMTS, GSM, CDMA, SS7, VoIP (H.323, H.224, H.248, etc.), and any other technology used in the access network. If your monitoring system can support all of these areas, you are well prepared for IMS implementation and should be able to manage your IMS network very closely.

Interfacing to the IMS

In an ideal world, connecting to the IMS would be as simple as turning on our communicator and selecting the media and service we wanted. Or better, simply communicating in whatever form tickles our fancy at any moment in time and letting the device control the rest. But alas, we are not in an ideal world.

Instead we are in a world filled with old and new, legacy and cutting-edge technology with new-fangled devices providing a myriad of applications. Yet despite all this new-fangled technology, we still have rural areas with multiparty lines and outside plant that most likely outdates most of those who maintain it. This means that implementing IMS architecture is not as simple as installing a few pieces of equipment and plugging them in. There is a lot of interworking that will be required as with any new technology.

Despite what many of the non-technical "marketers" may be trumpeting, the IMS will not replace existing infrastructure overnight. We will not wake up one day and suddenly be able to use our mobile phones as powerful multimedia devices connecting us to anyone anywhere in the world at the touch of a button.

Instead, we will find ourselves relying on the same devices we use right now, with some new features and capabilities gradually (year by year) entering our daily lives. Operators will transition their networks to IMS rather than simply replace everything from scratch. There are numerous factors for this.

First of all, there is simply too much investment in all that equipment out there. This equipment has not even paid for itself yet (at least not in accounting terms). Most operators are not willing to write their investments off for at least 10 to 15 years. Second, there is no business case to simply start over with IMS. IMS does not provide any new services or killer applications.

Rather, IMS enables new services and applications by providing the infrastructure that provides more robust control and management than the Internet does today. In the meantime, existing networks will have to interface to the IMS as part of a transitional strategy.

This is not uncommon in this industry. Many new technologies have been implemented through transitional strategies (the IN and SS7 is a good example of this). In fact we can draw many similarities and lessons learned from the early implementations of the IN.

The concept of an Intelligent Network (IN) using SS7 for call control started in 1964, continued to evolve as a standard throughout the 1970s, and was finally implemented in the U.S. in the 1980s. Europe adopted SS7 for connecting switches to one another for simple call control, but using a mesh network configuration without signal transfer points (STPs). The U.S. did the opposite, implementing SS7 with STPs for the support of 8xx service.

This was a long road from the inception of the technology to the actual implementation. But this is not the only example. ATM development began in 1969 but did not really begin taking off until the 1990s. This represents about a 20-year cycle, from beginning, to maturing the standards, to actual implementation.

Even newer technologies and architectures such as Voice over IP (VoIP) will require interworking with IMS, as these two architectures are not 100 percent interoperable with one another. Gateway functions are required to ensure this interoperability, which is the focus of the next section.

Circuit-Switched Domains (CS)

The circuit-switched domain, as shown in Figure 2.1, consists of analog and digital switching nodes, and while the voice may be digitized, it is not packetized. This requires additional nodes before the voice traffic (or bearer traffic) can be routed over an IP network. Packetizing the voice is only one requirement though.

Since the facilities within the circuit-switched domain have a different set of attributes than the domain itself, the method of signaling in the circuit-switched domain is not compatible with the packet network. In the circuit-switched world, Signaling System

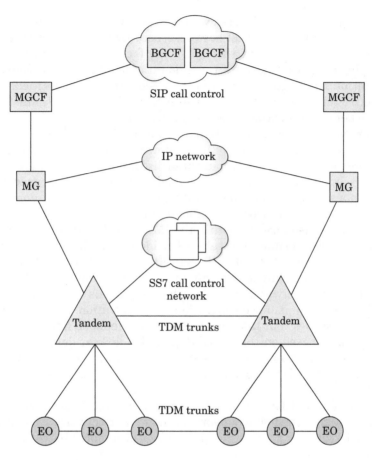

Figure 2.1 Circuit-switched domain

#7 (SS7) is used to communicate from one switch to another, the facility requirements for each call. The SS7 protocol specifically identifies the circuits to be used, the attributes of that circuit (such as codecs to be used), and even if echo cancellation is required (to name just a few identifiers).

More importantly, it is SS7 that allows switches to communicate with databases such as number portability to determine how to route calls through the network. SS7 and the IN architecture have enabled services such as wireless roaming, and many services we take for granted today (such as Calling Name and Freephone services).

SS7 can be divided into two functions. The first function is for the setup and teardown of voice circuits between switches in the network. This is the responsibility of the ISDN User Part (ISUP) protocol. The second function is for the connection and communication with various databases within the network. This is the job of the Transaction Control Application Part (TCAP).

This signaling method works very well within today's circuit-switched networks but is not compatible with packet networks, because the facilities have different attributes, and the media being sent through these facilities have different requirements. For example, in a packet network the amount of bandwidth is variable, and therefore the amount of bandwidth required for a session must be identified. The method used to packetize the voice (or whatever bearer traffic is being sent) must be identified so that the packet can be disassembled at the destination.

There is much more information exchanged, of course, in both domains, as we will see in the next few chapters as we break down the processes and procedures for the IMS signaling. Remember throughout these discussions that we are talking primarily about the signaling within each of these domains and not the bearer traffic.

When a subscriber initiates a call within the fixed circuit domain, that subscriber's location is identified through the facility he or she uses to gain access. This concept, by the way, is carried forward within the IMS as well, but with a major difference. In the circuit-switched domain, the subscriber identity is fixed to the facility (the telephone line) used to access the network. Since this location is fixed, mobility is not possible within the circuit-switched domain (at least not using today's traditional methods).

To access the network, the subscriber only needs to pick up the telephone and begin dialing. There is no consideration for registration or authentication; these are all manual processes that take place when the telephone line is ordered. The connection is then made based on the dialed digits. This is a far departure from the IMS model, where the subscriber can access the network from any location and facility, and the subscription tied to the facility itself is not usually the same as the subscriber accessing the network.

This is different for wireless, however. In a wireless network, the subscriber can be anyplace within the home network, or in another network. We will talk about the wireless model later, as there are many obvious differences between wireline and wireless challenges.

To route a circuit-switched call to the IMS, there needs to be interaction between the circuit-switched domain and the packet domain, and then the IMS domain. There may be cases where the circuit-switched domain connects directly into the IMS domain through gateways, but it is more likely that initially these connections will be made through VoIP connections.

The purpose of the VoIP gateways is to convert not only the bearer path to packet, but also the signaling required for routing the bearer traffic through the IP domain. VoIP uses a variety of signaling methods that are not always fully compatible with SS7, so breakout media gateway controller functions (BGCF) may be needed to convert this signaling. The purpose of the BGCF is to accept the SS7 signaling and use that signaling to determine how the bearer traffic is to be routed through the IP domain.

The BGCF does not necessarily "convert" in purest terms; it would be more accurate to say that the BGCF reads the SS7 signaling and uses the parameters within the SS7 signaling to determine where the bearer traffic needs to be sent. It is not converted, because most of the parameters are relevant to the circuit-switched facilities and do not apply to the IP domain. There are no echo cancellers, for example, within the

IP domain, so the parameters identifying these requirements are not needed within the IP signaling.

The BGCF will create new signaling for the bearer traffic within the IP domain, based on the destination of the bearer traffic and the attributes of the media themselves, but without regard to how it was received from the circuit-switched domain. This signaling then gets sent to the IMS access point, dependent on the architecture of the IP domain transiting the traffic. In other words, if the operator is using someone else's VoIP infrastructure for routing into the IMS domain, then the VoIP network will access the IMS through the P-CSCF within the IMS network.

If the operator is using its own BGCF, then the signaling may be sent to a P-CSCF within the same operator's network, or be sent to another operator's network acting as an IMS domain. There are also some cases where operators may implement two completely independent networks: one the legacy circuit-switched network and one the IMS network. They will need the BGCF then for generating the Session Initiation Protocol (SIP) signaling needed to communicate with the IMS, while still maintaining a connection back in the circuit-switched domain.

Consider the BGCF as the call controller between both the circuit-switched network and the IMS. The BGCF maintains the circuit-switched connection as well as the IP connections, for as long as the session requires. When the session is complete, the BGCF will then release the circuit-switched facilities by using the SS7 release procedures. It does the same on the IP side using the SIP procedures. This means that the BGCF must also be stateful, maintaining the status of every session under its control.

Once the BGCF creates the signaling using SIP, connection into the IMS is possible, but there are two variations of SIP. SIMPLE SIP is the first form developed for use in VoIP networks by the IETF, but it lacks many of the more robust attributes added by the 3GPP for use within the IMS. These extensions to the SIP protocol are required before a call can terminate within the IMS domain, because most of the extensions were added for authorization and authentication.

One other important note about routing calls from the circuit-switched domain into the IMS: the SS7 protocol is routed using Time Division Multiplexed (TDM) facilities, while the IMS is all IP-based. Many operators have already begun converting their backbone networks to all-IP. This includes the facilities traditionally used for routing of SS7 messages through the network.

Signaling gateways are used to convert the lower layers of the SS7 protocol to IP. The Message Transfer Part (MTP) is compatible only with TDM facilities and therefore must be replaced with a set of protocols compatible with IP facilities. SIGTRAN replaces the SS7 MTP while preserving the upper layers of the SS7 protocol (ISUP and TCAP). This allows operators to maintain their existing signaling infrastructure and transition the transport backbone to IP in preparation for an all-IP network.

The upper layers of the SS7 protocol (ISUP and TCAP) are still needed for managing the resources within the circuit-switched network, but they are not needed within the IP domain, as we have already discussed. The signaling gateway does not concern itself with these upper-layer protocols, leaving that responsibility to the BGCF/MGCF instead. So the operator can connect its existing SS7 network into the signaling gateway using IP as the transport behind it and begin routing all their SS7 signaling over the IP network.

So now that we have identified all of the pieces needed to connect from the circuit-switched domain, we can talk about the flow of traffic between the networks. The call is established within the circuit switch and routed through trunking facilities to a media gateway connecting into the circuit-switched domain. The MG is controlled by a MGCF, which means the signaling comes from the MGCF.

The MGCF then interfaces to the circuit switch for signaling via SS7. The MGCF may support circuit-switched facilities for interfacing into SS7, or there may be a signaling gateway function to convert the facility to IP (using SIGTRAN) prior to connecting into the MGCF.

The SS7 message is then used to determine what the bearer traffic is, what is required at the destination to access the bearer traffic, and the final destination for the session. The MGCF then creates a SIP message for routing into the IMS. The bearer traffic continues through media gateways to the final destination.

The MGCF interfaces into the IMS via a BGCF (the same function, but the BGCF acts as a signaling gateway into another network). The MGCF provides the SIP signaling needed by the IMS network to be able to set up a session within the IMS domain and control that session. The BGCF in turn connects into the P-CSCF for the serving IMS network, at which point the call is now under the control of the IMS network (at least the portion of the call within the IMS domain).

The Media Gateway (MG)   We have so far talked about the signaling traffic, but the bearer traffic must be routed as well. It is important to understand that the IMS itself is really about call control, and not bearer traffic. In other words, to control the transmission of voice, data, messaging, video, and any other media we can think of through an IP network, while maintaining security and preventing unauthorized access, there needs to be a set of functions where this can be accomplished. It has already been proven through decades of network experience that the best means for maintaining session control in any network is through a central core function, which is what the IMS was designed around.

The bearer traffic in this scenario is coming from a circuit switch which only connects via non-IP facilities. This bearer traffic must then be converted to an IP facility. This is accomplished by sending the bearer traffic to a media gateway (MG). The MG then connects to the legacy facility on one side, and IP on the other. As shown in Figure 2.2, the MG is under the control of the MGCF, which is the same as the BGCF, with the BGCF providing a gateway function into each individual network.

There will usually be multiple media gateways connecting to one single MGCF, providing a very economical means for supporting voice services in an IP-based domain. The media gateway also supports customer premises equipment by providing an IP facility to the customer network, and connecting via IP PBX or some other VoIP entity at the customer premises.

The media gateway is really the switching fabric in VoIP networks. The switching fabric is what interconnects one circuit (or port) to another for phone calls and other

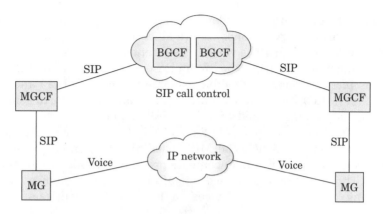

Figure 2.2 The media gateway (MG) and the media gateway controller (MGC)

session types. The media gateway performs the task of connecting the various circuits to one another for an IP network, under the control of the MGCF.

Once the signaling reaches the IMS domain, session control is managed by the IMS, providing the necessary call control signaling back into the circuit-switched domain. For example, a voice call may be originated within the circuit domain, passed to the MG and associated MGCF, and routed into the IMS. The bearer path remains within the IP network, under the control of the domain it is residing in.

If the session is terminated within the IMS domain, the SIP will terminate the call using the BYE method. This is then forwarded back into the circuit-switched network via the BGCF, which creates an SS7 REL message back to the switch.

The easiest way to relate to this is to remember that signaling is used to communicate from one network element to another regarding the facilities that are needed to form a connection end-to-end. If the facility is IP, then MGCF is needed outside of the IMS, and CSCF is used within the IMS. We will discuss more about the CSCF functions in later chapters when we discuss the inner workings of the IMS network itself. For now, know that the CSCF provides session control within the IMS domain.

The MG therefore becomes an important element when converting voice (or other non-IP media) into packetized media for routing through a packet network. But what if you are using a SIP phone within an IP network? The SIP phone acts as the SIP user agent, and while the MGCF is still used for call control within the VoIP domain, this can later be eliminated and replaced with the CSCF within the IMS.

The long-term goal is to have all SIP-enabled devices, eliminating the need for media gateways and other devices to packetize various non-IP media. Once a phone or other device is capable of packetizing the media and creating SIP signaling, the CSCF is all that is needed for session control.

The Media Resource Function (MRF) In the circuit-switched domain, service tones and recordings are all provided via the switch and other external elements (such as voice response units). These are not compatible with packet networks, however. Since there are no switches in the packet domain, there must not only be another function to generate the various service tones we are all accustomed to hearing, there must also be a controller for managing the various tones and recordings.

This function lies within the Media Resource Function (MRF). The MRF consists of two distinct functions: the media resource processor (MRP) and the media resource controller (MRC). The MRC can interface to multiple MRPs to generate all of the tones and announcements needed within the IMS domain.

The MRP then is responsible for creating the tone or announcement based on instructions from the MRC. The MRC uses SIP for communicating the requirements (such as what tones or announcements are to be provided). This replaces the methods used within the circuit-switched domain of routing calls over dedicated facilities to recording or announcement devices. In this case the facilities are IP connections.

The Media Gateway Control Function (MGCF)   The MGCF is widely used within VoIP networks already, and its role within the IMS is the same. The purpose of the MGCF is to provide connectivity at the control layer into the IMS, controlling all of the media gateways that are managing the bearer traffic.

The MGCF is also where some features and applications reside. In the circuit-switched domain, the switches provided many features and applications such as conference calling and call park. These are now under the MGCF, which is the "brains" of the operation.

The signaling function at the MGCF provides call control for all packetized voice as it enters into the network. The MGCF only communicates with media gateways within the same domain, however, and is not intended for interfacing with other MGCFs in other networks. The MGCF must then interface with a gateway function for interconnection with other domains.

The Breakout Gateway Control Function (BGCF) provides this functionality and interfaces with other BGCFs in other network domains. The MGCF then provides the call control within its own network; it may be implemented regionally or within the core network. Actual implementation of the MGCF function depends largely on the network topology and size.

For very large networks, it is more efficient to distribute the MGCF within regions, providing better control from a geographic perspective over growth and traffic grooming. This could be thought of as analogous with the tandem function within the circuit-switched domain, but it is just as analogous as the bridging function within LAN/WAN networks, although the traffic mixes will be significantly different.

Packet data networks often engineer bridges using the 80/20 rule. They are configured in such a manner that the bridge routes 20 percent of the traffic off-net, and 80 percent on-net. Tandems, on the other hand, are specifically designed for interconnection of different networks and network service areas.

The MGCF can be engineered in such a way as to route traffic between media gateways within its own control, and a smaller percentage of traffic routing to other media gateways outside of its control. They can also be implemented as more of a central function providing a sort of tandem function within a network.

Either way, their function within the IMS is to manage the packetized voice that is routed into the IMS domain, and to pass the control of these calls through the MGCF up to the Call Session Control Function (CSCF). They are needed only when interconnecting with the Public Switched Telephone Network (PSTN).

The Breakout Gateway Control Function (BGCF)   We have talked about the BGCF some already. The BGCF sits at the border of the network and controls interconnections between two networks. This means that the BGCF must provide some security capabilities to prevent unauthorized access into the network.

As shown in Figure 2.3, the BGCF interfaces with the various MGCF entities within its own domain and then provides the connection to another network connecting into the BGCF of that network. This allows operators to screen some aspects of the session information from competitors. For example, the operator may wish to "hide" the addresses of other nodes within its network. This would prevent the other networks from identifying the network topology.

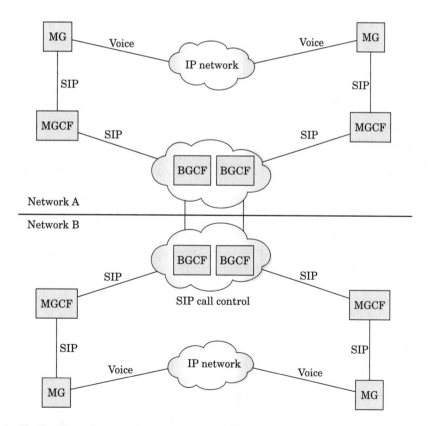

Figure 2.3   The Breakout Gateway Control Function (BGCF)

The BGCF can also provide encryption for all outbound traffic. This is typically achieved through the use of IPsec, where the message is encrypted and then placed inside another packet for routing. The message is then sent from one network to the destination network, encrypted and embedded within a normal signaling message. Also referred to as tunneling, this prevents intermediary networks from discovering many aspects of the network as well as the sessions themselves.

There is one caveat to tunneling: if you are monitoring traffic end-to-end within the network, unless there is some means of capturing the cipher keys, you will be unable to decrypt the message and monitor the traffic effectively.

In the IMS domain, it should be noted that the S-CSCF connects directly with the BGCF for outbound signaling. If the S-CSCF has determined that a SIP message is to be routed to another network, then the S-CSCF routes the SIP message directly to the BGCF.

If the message is to be routed internally (within the same domain as the S-CSCF), then the S-CSCF routes the message to the MGCF. This may be done directly or through another CSCF, depending on the operators' deployment and implementation.

Packet-Switched Domains (PS)

One would think that within the packet domain, there is no need for gateways or other entities to route into the IMS. However, the packet domain is not necessarily IMS compatible. In today's networks, operators have been evolving the network from circuit-switched to packet-switched, which means you have hybrid networks.

A good example of this is the GSM network. GSM started as an IN/SS7 architecture model, using the SS7 protocol to communicate to internal databases that manage subscriber location and registration within the network. The switches themselves use circuit switching when routing calls between networks, but in many cases X.25 packet switching was implemented for calls within a base station subsystem, and back to the switch itself.

As GSM matured, it began evolving certain aspects to packet-switched using IP. This is indeed how IMS came about in the first place, as part of this evolution. But there are still legacy components that must be supported, until such a time when handsets are equipped as SIP User Agents (UAs) and can communicate directly with the IMS infrastructure without interfacing with any other entities.

The packet-switched domain can be wireline focused or wireless focused, so this section is broken into two parts to discuss both Voice over IP (VoIP) and wireless implementations.

VoIP Domain

We have talked quite a bit about the VoIP domain already in the sense that calls from the circuit-switched domain must then be routed through the VoIP domain to packetize the voice, and to convert the signaling from SS7 to SIP. This being the case, we have already identified several of the functions within this domain, but we will cover them again here from a different perspective.

The VoIP domain consists of entities that provide the functions commonly found in a telephone switch, with one major difference: the functions are distributed. When the concept of VoIP first began, the idea was to separate the function within a switch and provide these functions in a more economical fashion. As shown in Figure 2.4, the switch can be broken into three major functions:

- Switching fabric (or matrix in legacy terms)
- Signaling
- Applications

The switching fabric is the cheapest part of the switch. This is where circuits are connected to other circuits within the switch, for connecting of a call. Think of the telephone operators of days far gone and how they used their answering positions to connect calls with other parties. The switching fabric does this same function under the control of a processor (in other words, we automated the operator function).

Detach this function from the processor, and you have a relatively inexpensive device that can be placed throughout the network. For example, an operator providing service to a remote rural area cannot afford to deploy a full-blown circuit switch into this market, but it can deploy an inexpensive media gateway in the market and place the controller someplace more central, supporting media gateways in many different rural areas.

So now you have the switching matrix being provided by the media gateway. The media gateway does not have any processing power, and therefore it must rely on a

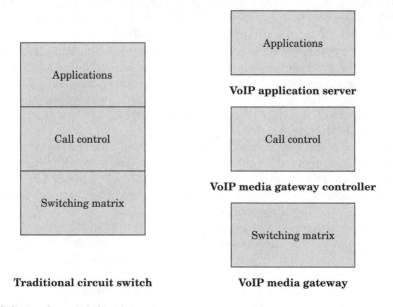

Figure 2.4   Splitting the switch functions

controller to tell it how connections are to be made, and to control those connections. The media gateways do not have any signaling between themselves per se; they communicate with their controller. A number of protocols are used between the media gateway and the controllers. This is another issue with many VoIP implementations, because it is difficult to interwork many different brands of devices if they do not support the right protocols.

The controller is a more central function. The media gateway controller can sit within the core network, or it can be placed within a specific geographical location and provide call control for many different media gateways. The media gateway controller function (MGCF) is the brains of the VoIP network. The MGCF is also where the signaling portion of the switch is placed. It is the MGCF that communicates with other MGCFs within its own network, providing call control between there various entities.

Since the processing for a call is not distributed in key areas, this function can be used to support multiple communities of interest, rather than deploying one per town that can only be used to support that one town. This is the direction of VoIP deployments and the major attraction from an operator's perspective. It does not come without issues, however.

The multitude of signaling protocols is one of the largest issues and has caused implementation nightmares for many operators. This was especially true for wireless operators trying to support applications. Since wireless is heavily dependent on databases (the Visited Location Register and the Home Location Register, for example), it becomes paramount that every entity within the network be able to communicate with the Mobile Switching Center (MSC) to access these resources.

The applications and features in the VoIP model are also separated from the switching function and deployed on application servers. This presents many cost savings and allows operators to implement just about any type of service they want to offer, as long as their MGCFs can communicate and support it.

A voice application server, for example, may provide mobility services, conferencing services, or even messaging. Operators such as Skype and Vonage are heavily dependent on these application servers because this is where mobility is supported. Mobility, by the way, is the concept of receiving calls no matter where you are located. The network always knows your IP address and always knows how to route calls to your device (as well as what device to route calls to).

In essence, the VoIP model is almost IMS ready, except that many different protocols are used to support all of the communications between all of the network entities, and very few security and authentication controls are provided by the network. Almost all of the security and authentication controls are functions of the MGCF and require implementation by the operator.

However, the concept of inserting a module within the subscriber device for the exchange of authentication credentials is largely a wireless concept. VoIP implementations may have some security, but there is no means of authenticating the subscriber device today.

Certainly an operator can provide packetized services today, with all of the features and capabilities that can be provided within the IMS, and indeed many are doing this now. However, they are lacking in the control and security area, which is why IMS has become so important to so many large operators.

The VoIP domain, then, still must have a connection into the IMS, which is through the BGCF/MGCF. If the operator owns both a VoIP network and an IMS network, then it can pass calls/sessions between the two domains using the MGCF. However, if the VoIP provider is not the same as the IMS provider, then the BGCF is used as an interconnect part. As you will learn in later chapters, there are specific functions provided by the BGCF that differentiate it from the MGCF.

Once the signaling reaches the BGCF, it must be routed to the P-CSCF within the area (or whatever means the operator chooses for P-CSCF assignment). Remember that all sessions must be routed through the P-CSCF first, which then routes to the I-CSCF, which is then responsible for routing to the S-CSCF.

VoIP only supports voice and data, and therefore there must be other interfaces to support other media types, for example, the delivery of messaging in a wireless network and the support for video over IP (IPTV). These require other network elements and a separate means for delivery. This is because, even though the network itself cannot support the media types, the network elements are designed for voice only.

Support of these media types involves overlay networks, such as GPRS in the GSM network. GPRS is explained in the next section, but it provides an overlay network supporting various data services such as Internet access, video and audio support, and other packetized services.

General Packet Radio Service (GPRS)

The GPRS network works as an overlay to the existing GSM network. All packet data received from a GSM handset is routed from the base station controller (BSC) to the Serving GPRS Support Node (SGSN). The SGSN is a packet data node supporting multiple media types in a packet network.

As shown in Figure 2.5, the SGSN provides connectivity to other SGSNs within the same network, acting as the packet data network for the wireless operator. None of this traffic is routed through the Mobile Switching Center (MSC), since this switch is designed to support voice.

To interface to external packet networks, the Gateway GPRS Support Node (GGSN) is used. The GGSN provides connectivity to other packet networks, including other wireless carriers' GPRS networks. This overlay network then provides complete packet data support within a wireless environment within the core network, but this does not extend to the air interface. Other technologies exist to provide support of packetized broadband services at the air interface.

The GPRS network provides and manages connections to packet networks. This means that GPRS must provide some fashion of session control for each of the connections within its domain. When connecting to the SGSN, a Packet Data Protocol (PDP) *context* is created. Think of the PDP context as the connection into the packet network, each connection being identified by its own unique PDP context.

When connecting from a wireless device to an application server in the IMS domain, the SGSN/GGSN creates a PDP context for the session and then manages this connection

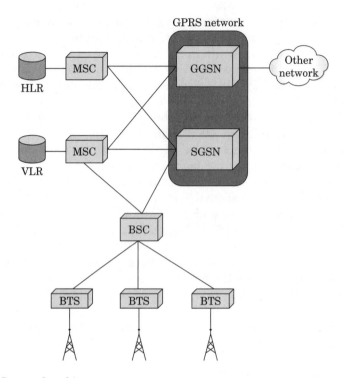

Figure 2.5 GPRS network architecture

on the GPRS side. The GGSN then interfaces with the P-CSCF, which provides the session control within the IMS domain. When connecting to servers and applications, the GPRS network uses the Access Point Name (APN) for the device to identify which services it is accessing (based on the server name being connected to).

UMTS and CDMA Domains

The Universal Mobile Telecommunications Service (UMTS) provides broadband packet services enabling wireless subscribers to use Internet services and other multimedia-type services such as video and multimedia messaging. With a bandwidth of 2 MBps, UMTS provides the bandwidth required to support some of these services that require more bandwidth, such as video (IPTV). Voice is also supported through this interface but is routed from the UMTS nodes to the MSC (the circuit-switched portion of the network).

As you can see in Figure 2.6, accessing the IMS through a UMTS interface (or any other access technology for that matter) within a GSM network still involves routing the packet data through the GPRS SGSN/GGSN nodes. The Radio Network Controller (RNC) in the UMTS portion of the network interfaces with the SGSN within the same network, which in turn interfaces with the GGSN.

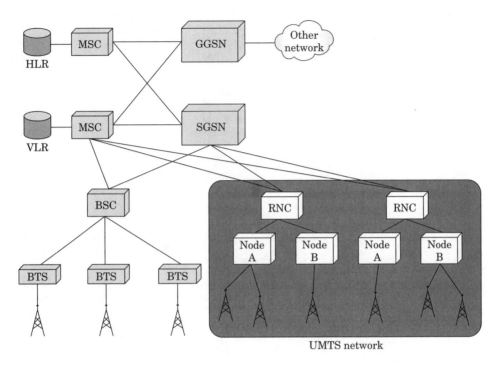

Figure 2.6   UMTS network

It is the GGSN that then provides the connectivity back into the IMS P-CSCF. The P-CSCF is always the first point of access in the IMS, regardless of the access method. Voice traffic does not necessarily go through this same route, because (as you'll recall) the voice must go through media gateway, under the control of the MGCF. SIP signaling is the only data that is routed through the CSCF functions of the IMS.

This is the same then as in wireline VoIP, where the voice is sent to a media gateway (MG) for converting to packet, and the signaling is sent to the MGCF for conversion to SIP. Once the signaling has been converted to SIP, the MGCF is then able to communicate directly with the IMS P-CSCF.

CDMA is a little different in that the architecture does not include the SGSN/GGSN (see Figure 2.7). However, there is an equivalent function: the Packet Control Function (PCF) interfacing with the Base Station Controller (BSC), which is responsible for the routing of packet data to the packet network within CDMA.

The Packet Data Service Node (PDSN) then provides connectivity to packet services such as the Internet, or other packet networks. The PDSN therefore serves the same role as the GGSN, acting as the gateway into other networks. The PDSN interfaces with the media gateway within the IMS domain if there is voice; otherwise, for packet data the interface is to packet entities within the packet domain. The session is controlled by SIP, so the SIP signaling is routed to the P-CSCF within the IMS.

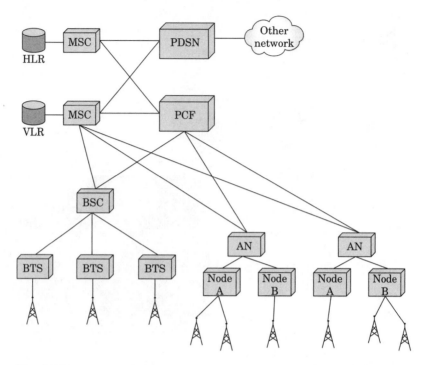

Figure 2.7   The CDMA network and packet domain

From IMS to IMS

In a perfect IMS universe, there is no need for access networks to support anything other than IP and SIP. Of course, we are not in a perfect IMS world (at least not yet), and there is a significant amount of network still in service that is not packet. For this reason, we will not realize the full benefits of a total SIP network for a long time.

When I say total SIP network, I am referring to a network where all of the devices are equipped with SIP user agents and are capable of supporting IMS SIP. This would allow the devices to interface with the CSCF within the IMS domain directly without any conversions or signaling gateways (see Figure 2.8). MGCFs would not be required since the devices are already SIP capable.

These same devices would be able to transmit everything in packet mode, eliminating the need for media gateways and other support devices in the network. Regardless of the media type, all transmissions would be packet data, and signaling would be SIP. This is the future direction of IMS.

This is not as far off as some might think. It is possible we may see implementations of SIP-enabled handsets in the very near future, which will only serve as a major incentive for operators to move rapidly to an all-IMS environment. With support for packet data and SIP network-wide, many elements and systems can be completely eliminated from the network.

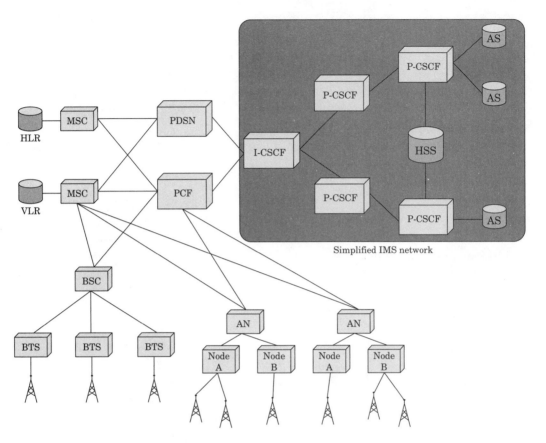

Figure 2.8 The pure IMS network

Of course, this is still talking about an ideal world. There are many back office support systems that will have to transition, because it is simply not possible to cut the cord overnight and toss these systems out of the window. The entire business of the operator is based on many of these support systems, so throwing them away is out of the question.

Likewise development is still lacking for many of these systems that support IMS. For example, the entire area of billing is still being defined and refined. Operational support systems (OSSs) are moving in this direction quickly, yet they have a different view today than what operators may need going forward as their business models change. This is yet another reason why we may not see full SIP-to-SIP, IMS-to-IMS interoperability without gateways and other devices to interface to legacy platforms and systems we have already discussed.

This is truly an evolutionary approach to IMS implementation, and for any operator who already has a legacy network in place, it is the only approach that makes any business sense. The long-term goal is to support SIP-enabled devices in a pure IMS environment, but the road to get there will be long and require transitional approaches.

3

Session Initiation Protocol (SIP)

The 3GPP had many protocols to choose from when creating implementation standards for VoIP. There already existed many different call control protocols, such as MEGACO, Media Gateway Control Protocol (MGCP), H.323 protocols, and many more. This, in fact, is one of the problems when trying to implement VoIP networks; there are many different protocols that have to interoperate, yet there lacks an implementation and interoperability standard for VoIP.

The SIP protocol was actually derived from existing protocols developed to support other forms of media. Both the Simple Mail Transport Protocol (SMTP) and the Hypertext Transport Protocol (HTTP) were used to create SIP.

SIP was chosen because while it may not have been the perfect protocol, it did support all forms of media, and with some modifications it could be made much more robust and secure. The 3GPP has defined many extensions to SIP through the IETF for this very reason.

The SIP protocol within the IMS is used to control everything that a subscriber does. This includes voice, messaging, e-mail, and data transfers. This is a big advantage over conventional VoIP implementations because it allows the operator to support all media types using one common protocol. It also means the operator can invest in one set of back office applications supporting SIP rather than having to invest in many different systems supporting disparate protocols.

The use of the SIP protocol for all things IMS simplifies the implementation of an all-IP network considerably for many different reasons, including those already discussed. Yet SIP is not the perfect answer, as there are still things missing from SIP that are needed in the IMS. These are the extensions just mentioned.

SIP is not a new protocol. In fact, SIP was created from two well-known protocols widely used on the Internet today: Simple Mail Transport Protocol (SMTP) and Hypertext Transport Protocol (HTTP). You will see many similarities between these protocols and SIP, including many of the procedures defined for mail delivery and Web browsing. However, there is also a caveat to using these protocols.

Internet protocols are not well known for robust security. The Internet community leaves security to the end devices that access the network. This leads the service providers providing Internet access to ignore breaches and other security holes in their networks. It likewise means that the consumers must take things into their own hands to ensure they are not violated through a lack of security on their computers and access devices.

This model does not work for a traditional telecommunications service provider. With revenues tied directly to access and usage of the network, as well as many services, operators cannot afford security breaches stealing these very same assets. For this reason telecommunication networks build very robust security procedures to prevent fraudulent access and denial of service attacks.

The requirements for interoperability and security are therefore much higher in these networks than they are in conventional Internet networks. This is the reason the 3GPP set out to define a version of SIP that would support the authentication and authorization procedures needed for wireless networks. The many extensions that have been defined by 3GPP were added for exactly this reason, as well as charging of services within an IMS.

Needless to say, SIP as it has been redefined for use within the IMS adds much better security to the network than a simple VoIP deployment would (or could). This is the primary reason that any service provider that is moving to an IP-based network should seriously consider SIP and the IMS as its best means of success. The business case is sometimes more difficult when it is risk-based, but it is the best business case there is for deploying IP in the network. Anything less will leave the service provider with serious vulnerability issues placing revenues at risk.

This chapter is by no means an exhaustive description of SIP, nor is it intended to provide all of the details according to the IETF RFC 3261. That is covered in another book, *Session Initiation Protocol (SIP): Controlling Convergent Networks* (McGraw-Hill, forthcoming). The purpose of this chapter is to provide an overview of the SIP protocol as it relates to the IMS.

SIP Protocol Structure

The SIP protocol comes in three parts:

- Start/status line
- Message headers
- Message body

The start line is the first line within a SIP request message. It contains an address referred to as the Request-URI, as well as other information regarding the version of SIP creating the message and the SIP method (think of the method as the message type). Here is an example of a standard request:

```
INVITE  SIP:travis.russell@tekelec.com  SIP/2.0
VIA: SIP/2.0/UDP pchome101@aol.com:5060; branch=z9hG4bK74gh5
```

```
FROM: Deby Russell <sip:deby.russell@aol.com>;tag=9hz34567sl
TO: Travis Russell <sip:travis.russell@tekelec.com>
MAX-FORWARDS: 70
CALL-ID: 82167534@126.18.27.0
CSEQ: 1 INVITE
CONTACT: Deby Russell <sip:deby@126.18.27.0>
CONTENT-TYPE: application/SDP
CONTENT-LENGTH: 154
```

Note that the first line contains the type of request *(INVITE)* with the address where the request is to be sent. This address may be simply the next hop in the network, or it may be the final destination, depending on how routing is configured within the network. The SIP/2.0 indicates the version of SIP that was used to create the request. This allows receiving nodes to determine how the message is to be processed.

The rest of the lines consist of headers. Each of the headers contains expected parameters providing additional details about the request. The last line of the message is the CONTENT-LENGTH, which defines the length of the message body itself. The message body then follows.

The contents of the message body are described in the header CONTENT-TYPE. In the preceding example, the content is the application "Session Description Protocol (SDP)." This means that the message body will contain yet another protocol known as the Session Description Protocol (SDP) that describes the details about the session being requested.

So as a general rule, if a "dialog" is needed between two devices, the message body will contain requirements for a session. Think of a session as a "connection" between the calling party and the called party. If there is no "connection" needed and the intent is to simply deliver content (such as a text message), then the message body can be used to contain the actual content to be delivered.

If the message is a response, the first line is referred to as the status line. The status line provides a status code indicating the status of the request. The status code falls into one of six categories, as detailed in another section of this chapter. Below is an example of a standard response message.

```
SIP/2.0  180 RINGING
VIA: SIP/2.0/UDP  128.10.10.1  raleighproxy.com:5060; branch=z9hG4bK63x2f
FROM: Deby Russell <sip:deby.russell@aol.com>;tag=9hz34567sl
TO: Travis Russell <sip:travis.russell@tekelec.com>
MAX-FORWARDS: 70
CALL-ID: 82167534@126.18.27.0
CSEQ: 1 INVITE
CONTACT: Deby Russell <sip:deby@126.18.27.0>
CONTENT-TYPE: application/SDP
CONTENT-LENGTH: 154
```

Note in this example that the first line indicates the status of the request. In this example the provisional response 180 RINGING is given to indicate that the network is attempting to alert the called party. The status line is then followed by the various headers as necessary to support the response and any other dialog between the two entities.

Basically, the status code can be provisional (in progress), successful, redirect, or one of three failure response categories. Within each of these categories there are a number of predefined codes.

There may also be headers containing other parameters below the status line. The headers follow directly after the status line and the start line, and before the message body. All SIP messages are in text format (rather than binary coded); therefore, they are very simple to decode. Likewise, the various headers and parameters are also very easy to understand.

The message body can contain content, or it can contain details about the bearer traffic being sent in a different message (and protocol such as Real Time Protocol, or RTP). The message body for a voice call, for example, would contain the Session Description Protocol (SDP), describing the voice content itself (such as the encoding used and any other special requirements).

Note that a SIP message does not necessarily have to have session description content contained within the message body. The message body can be used to transport anything, in addition to a session description. For example, text messaging is carried through the network using the message body of a SIP message. This simplifies the network requirements to support messaging significantly (and is also analogous with SS7, where text messages are often carried in the Transaction Capabilities Application Part, or TCAP).

SIP is flexible and provides a lot of opportunity for operators looking to simplify their networks while enabling interoperability and security. This is the spirit of the 3GPP specifications for the IMS, and why SIP was chosen for session control. The structure is also simple, and it allows for extensions to be defined by organizations looking to enhance the capabilities of the SIP protocol.

Indeed the 3GPP has done just exactly that by submitting protocol extensions to the IETF to support the various functions defined by the 3GPP for supporting the IMS. Charging in the IMS is a good example of protocol extensions. Security is another good example where the protocol has been expanded through protocol extensions. All of these extensions are discussed throughout the book when we discuss the various procedures used throughout the IMS.

SIP Methods

SIP methods can be thought of as message types. They identify the request being made by the user device (or the network entity in some cases). As we will discuss later, these requests are used both by the destinations to determine what action is being requested and by the intermediary network elements (such as the Call Session Control Function) to determine how the message should be routed through the network, and in some cases how the message should be treated prior to delivery to the end destination.

There are several basic SIP methods currently defined for use in the IMS:

- Ack
- Bye
- Cancel

- Info
- Invite
- Message
- Notify
- Options
- Register
- Subscribe
- Update

Ack When an *INVITE* has been sent, the sender awaits a response by the destination. However, a dialog is not established between the two entities until the originator of the *INVITE* sends the *ACK* method. This method is the final handshake required to establish the dialog and allow for the session to begin.

If a device receives multiple 2*xx* responses to an *INVITE* (as would be the case for a conference call, for example), it must then generate an *ACK* for each of the responses. The *ACK* contains the same credentials as the *INVITE* and may also contain a message body carrying SDP or other content.

The details of how *ACK* is used to establish a connection are provided in Chapter 5; there you will find details about forming a dialog and the handshake sequence required between the two devices.

Bye This is the simplest of methods. To release a session in progress, either device sends the *BYE* method to release the session. It can be initiated by either end of the session. Like all other methods used during a dialog, the *BYE* does require a response and acknowledgment prior to the session being released.

Cancel This method is used by a device when it wishes to cancel a request prior to receiving a response from the destination. For example, if a cell phone is attempting to set up a call and sends an *INVITE,* but the user then hangs up their phone immediately after dialing, the call is released and the session is ended. However, the session was never really established, because a response was never received by the originator.

When these scenarios happen, the device will send a *CANCEL* method to the destination, which then cancels the original request and discards the request message. What happens if the *CANCEL* is received by the destination after the response was sent (they cross each other)? When the response is received by the originating device, it is ignored.

Info The *INFO* method was added as an extension to the SIP protocol for the purpose of communicating mid-call information between endpoints. Information could include dialed digits, possibly a picture file being exchanged between the two callers, or even ISUP signaling parameters that one of the endpoints needs to process to support the call in progress.

The RFC does not define exact uses for the method, other than to outline the possible uses for this method and how it should be processed by the various entities within the network. Since the contents of the *INFO* method may contain information that needs to be processed by entities supporting the call, it must be routed using the same path as the call setup. This will ensure that the method is received by any proxies along the way.

The actual content of this method could be identified in a message header, in which case the parameters within the message header would be processed by the various proxies in the path accordingly. As an alternative, the content could also be carried in the message body, as would be the case for a picture file.

If the content is in the message body, then the endpoint must know how to handle the message body. This obviously would be identified in the *CONTENT-TYPE* header. One of the options could simply be to render the content to the display of the subscriber's device (again as would be the case for a picture file).

Invite   The next most used method is the *INVITE*. This is used to establish a session within the IMS. Think of the *INVITE* as an invitation to another user to join in a conversation (or an e-mail, or an instant message, or whatever type of session is being set up).

The *INVITE,* of course, must then carry details about the session being established, as outlined in Chapter 5. The *INVITE* can also be sent to application servers providing services such as Blackberry service (for e-mail) or Web browsing. Anytime a device needs to establish a connection (virtual, of course), the *INVITE* is the method used to establish that connection.

Message   The *MESSAGE* method was added by the 3GPP to support text messaging in the IMS. Rather than send an *INVITE,* which was developed to establish a session, a new method was created for the simple task of delivering a text message.

The *MESSAGE* method is also used for sending instant messages through the IMS, but only when the IM is being sent in "page" mode. In other words, the IM is being sent in connectionless mode without establishing a session with the destination. If connection-oriented service is required, then an *INVITE* is used to establish a dialog between the sender and receiver.

The content of the IM (the text itself) is carried in the message body of the *MESSAGE* method. Again, this is only when the IM is being sent using a connectionless service. When a session is established using the *INVITE* method, the text itself is placed in a Message Session Relay Protocol (MSRP) after the session is established.

Notify   *NOTIFY* is used by the S-CSCF to notify application servers (or any other entity) that have "subscribed" to event notification (registration updates) that a subscriber has changed his or her registration. The *NOTIFY* method will contain the changes that were made through the new registration.

One example of how this method is used is message waiting. A message waiting server would "*SUBSCRIBE*" to event notification. When a message was left on the subscriber's voicemail, the message waiting server would then send a *NOTIFY* method to the subscriber device, with the appropriate header identifying the type of event that has occurred. The SIP device would then have to determine how to process the *NOTIFY* and what action should be taken based on its own capabilities and applications.

Options   The *OPTIONS* method is used to query application servers about their capabilities. This allows the network to ensure that a particular server is able to support requested services prior to establishing a session with that server.

It can also be used to query another user device about its capabilities prior to sending an *INVITE* to the device. This way the requesting device can determine if the endpoint can even support the type of session it is about to establish without sending the actual request.

The response to the *OPTION* method contains the *ACCEPT* header, containing the methods and capabilities supported by the responding entity. For example, if the device supports the SDP application, then the response would contain the *ACCEPT* header with `application/SDP` as its parameter.

Register   The *REGISTER* method is usually the first method to be initiated by a device right after the device is turned on. The purpose of the *REGISTER* is to notify the network of the device's location (IP address). This is so the network knows how to route messages to the device, so callers can find the subscriber wherever they are located.

This is one of the unique features of an IP network: mobility. Certainly this functionality can be provided through the Internet in general, but simply providing a location for a subscriber is not a secure means of providing a service.

For this reason the *REGISTER* messages also provide credentials that are embedded in the device itself. Only the network provider and the device know these credentials. To make service even more portable, a Universal Integrated Circuit Card (UICC) is used to store the credentials and is inserted into the device by the subscriber. This allows subscribers to use devices that they purchase from any store, without giving services away.

The credentials are provided using the *REGISTER* message only after the registrar rejects the first *REGISTER* message. This means that the *REGISTER* message content will not always be the same. In some cases the device only needs to update its location, while in other cases the device needs to register when it is activated. The procedures for registration are detailed in Chapter 5.

Subscribe   *SUBSCRIBE* is used by application servers to request updates from the S-CSCF and the HSS whenever a subscriber's registration changes. For example, a Presence server needs to know anytime a subscriber changes locations or activates another device in the network. The application server hosting the Presence application will send a *SUBSCRIBE* to the S-CSCF to notify it that it would like notification anytime a subscriber changes his or her registration.

The S-CSCF then uses the *NOTIFY* method to let the application server know of the changed registration, accompanied by the registration changes.

Update   When a device wishes to establish a session, it uses the INVITE method. To update the destination regarding changes to the session, the requestor would typically use the re-invite (sending an INVITE again using the same CALL-ID). The changes to the session are then contained in the new INVITE.

However, this also changes the dialog between the two devices, which would not be desirable if the initial INVITE had not yet been acknowledged. If an INVITE is sent, and an update to that INVITE is needed, then the UPDATE method can be used to send new information about the session without altering the dialog (which is yet to be established).

In essence this is a method used to notify the destination of session changes before the destination has a chance to accept the original session. To fully understand this concept, you must first understand the concept of a dialog, which is explained in full detail in Chapter 5. The dialog is not the same as the session but rather comprises communications between two devices *about* a session.

SIP Requests

In any communication, there is a request and a response. The same is true in SIP networks. We already discussed the concept of the request and response when we described the various methods used in SIP.

The request is a call to establish a dialog. A dialog is a virtual connection between two endpoints. A dialog cannot be started until the two endpoints have exchanged information regarding the parameters for the dialog and have acknowledged interest in participating in the dialog. This is accomplished through a handshake procedure.

For the purposes of this discussion, a handshake takes place when an endpoint sends a request to a destination, the destination agrees to the session via a successful response (200 OK), and the requestor returns an acknowledgment. Once this sequence of events has been completed, the dialog is complete. The handshake procedure is discussed in more detail in Chapter 5 when we talk about establishing a session in the IMS.

The method used determines the type of request being made. We have already discussed the different methods defined for SIP. Each of these methods is used for specific needs. For example, if content such as voice or video is to be exchanged using a connection-oriented-type service, then *INVITE* is the method used. However, if the message is carried within the body of the method itself, then perhaps the *MESSAGE* method is used.

The first line of a request is the start line, followed by the headers. Here is the format of the start line for a request:

```
METHOD  (space)  REQUEST URI  (space)  SIP VERSION  (crlf)
```

The request-URI is the address that the request is being sent to, either in the form of a URI or in the form of a TEL URI. The request-URI does not necessarily have to be the end destination. This is dependent on routing in the IMS. Usually the request-URI designates the next hop in the network, as identified in the *ROUTE* header. Details about routing are explained in Chapter 5 as well.

As discussed previously when we talked about the SIP message format, the headers contain parameters that then describe the request in more detail. These headers are identified in the following sections.

SIP Responses

No request can be completed without a response. There are numerous types of responses. Each of the responses falls into one of six classifications, identified by a preceding number. The first digit of the number identifies the class of response, while the following two digits identify the specific response being given. The response types are:

- 1*xx* Provisional
- 2*xx* Successful
- 3*xx* Redirect
- 4*xx* Client Failure
- 5*xx* Server Failure
- 6*xx* Global Failure

1*xx* Provisional Responses   When a request is made to establish a dialog, a timer is set within the device. When the timer expires, the request is sent again. This is to ensure that a request does not get lost in the network and the device simply waits forever. However, this can also create a lot of traffic within the SIP network if the timer expires quickly, since the device will simply repeat the request continually until a response is received.

For this reason, various network entities within the IMS may send a provisional response. The provisional response indicates that the network is delivering the request to the destination, and it may even provide some form of status.

This prevents expiry of the request timer and lets the device know that the request is being delivered. There are several provisional responses supported:

- **100 Trying**   Indicates that the network is attempting to reach the destination (sent by a proxy to prevent retransmission by the requestor).

- **180 Ringing**   This is sent by the request recipient to indicate the request has been received and the device is alerting the subscriber.

- **181 Call is being forwarded**   Indicates the call is being forwarded to another number.

- **182 Queued**   Indicates the called party is not available, and the receiver of the request is going to queue the request for later delivery.

- **183 Session Progress**   This is sent by intermediary entities to indicate the session request is being routed through the network and is being processed by the various entities. For example, a Media Gateway Controller (MGC) receives a request and immediately returns this response to indicate it is processing the request.

2*xx* Successful Responses   Success responses are sent by the receiver of a request once the receiver has agreed to establish a dialog with the requestor. This is the first part of the handshake required to establish a dialog. The following responses are defined:

- **200 OK**   This is sent to indicate that the receiver has accepted the request and the session can begin (once the *ACK* is sent by the requestor).

- **202 Accepted**   This is sent to indicate the request has been accepted but more processing is required prior to completing the request.

The *ACCEPTED* response is sent to notify the requestor that the request has been received but more processing is required prior to a dialog being established. The dialog is not established until the *ACK* method is sent by the requestor. This requires the 200 OK responses. The 202 ACCEPTED responses are typically used with instant messaging, for example.

3*xx* Redirect Responses   In SIP networks, a session can be redirected to another location. For example, if a request is sent to a cell phone that has been made unavailable, the profile for the user may dictate that calls be sent to another device, located elsewhere in the network. The redirect responses are then used to notify the requestor that the request is being redirected to another address. Here are the redirect responses currently defined:

- **300 Multiple Choices**   This is sent to the client to provide additional addresses for a subscriber as alternate destinations.

- **301 Moved Permanently**   The *CONTACT* header provides a new address for the subscriber, which is then used by the receiving device's software to update address books, etc.

- **302 Moved Temporarily**   This response also provides a new address, but the address is a temporary address; therefore, receiving devices will not use this information to update address books.

- **305 Use Proxy**   Indicates the address of a proxy that must be used to provide a requested service.

- **380 Alternative Service**   Indicates that the call was not successful but other services requested in the same session request can be provided.

The exact response provides the reason for the redirect. For example, the subscriber may have temporarily moved to another network or may have changed networks permanently. The URI that the session is being redirected to is provided as well.

4*xx* Client Failure Responses   When a device receives a request that it cannot accept because of an error made by the originator, then it will respond using the client failure responses. For example, a request sent to a device can be denied prior to a dialog being established by returning the 403 Forbidden response. Here are the defined client error messages:

- **401 Unauthorized**   This is sent by the S-CSCF to challenge a device when it first sends a *REGISTER*. The device will then send another *REGISTER* containing credentials per the registration process.

- **403 Forbidden**   This is used when a call is being rejected prior to a dialog being established. The dialog cannot be established until a 200 OK has been received as a response and the *ACK* has been sent as the final sequence in the handshake process. This response would be sent prior to the 200 OK.

- **405 Method Not Allowed**   This is sent when the recipient of a request recognizes a method used in the request but does not support it.

- **416 Unsupported URI Scheme**   The recipient of a request does not understand the URI format being used in the request.

- **420 Bad Extension**   The recipient of a request does not understand the *OPTION-TAGS* in the *PROXY-REQUIRE* header.

- **481 Call/Transaction Does Not Exist**   This is sent by the recipient of a *BYE* method that cannot be correlated to a call or session in progress.

- **483 Too Many Hops**   This is sent by stateful proxies when they receive a request and the *MAX-FORWARDS* header contains a value of 0.

- **486 Busy Here**   This is sent by the recipient of a request when they do not wish to accept the requested session. This might be seen, for example, when a request is sent to multiple devices and one of the devices is rejecting the request.

- **487 Request Terminated**   This is sent by the recipient of a request when the internal timer expires prior to a 200 OK response being sent.

- **488 Not Acceptable Here**   This is used to reject an offer contained within a request. The reason for rejecting the offer will be contained in the response.

- **491 Request Pending**   This is sent by the recipient of a request when that recipient is waiting for a response from the same requestor.

Remember that these are responses; therefore, they do not come from the client (originator of the request). Rather, they are used to indicate the session request is being denied because of a problem in the direction of the client.

5xx Server Failure Responses   Server failure responses are sent when a request is denied because of an error at the server (or destination of a request). Remember that these can come from a subscriber device or they can come from an application server providing a service to the subscriber (such as voicemail or video streaming). Following are the defined 5xx responses:

- **500 Server Internal Error**   This is sent, for example when messages are received out of order.

- **501 Not Implemented**   This is sent if the received request contains a method that is not recognized. For example, if a *MESSAGE* is received, but the recipient does not recognize this method, then this response is sent.

- **502 Bad Gateway**   This is sent when a downstream server sent an invalid response while the proxy was acting as a gateway

- **503 Server Unavailable**   This is sent by a proxy (such as CSCF) when the proxy is out-of-order or congested, and not able to process the request. The proxy may indicate a time when the request can be attempted again using the *RETRY-AFTER* header.

- **504 Server Time-out**   This is sent when waiting for a response from an entity in another network and the response timer expires.

- **505 Version Not Supported**   Indicates the server does not support the SIP version specified in the request.

- **513 Message Too Large**   This is sent when the received request was too large for the server to process.

6*xx* Global Failure Responses   The rest of the possible error scenarios are covered by the Global Failure responses. These responses are used when a session fails and there is no way to reach a subscriber at any of that subscriber's locations.

For example, a subscriber may have several devices registered in different networks. A call would typically be routed to the device addressed in the request, but the subscriber may have configured their services so that the request for a session is redirected to one of their other registered devices (e.g., call forwarding).

If all of the registered devices are unavailable, this would be considered a global failure because the subscriber cannot be reached at any location in the network. Following are the defined responses:

- **600 Busy Everywhere**   This is sent when the recipient of a request knows that a subscriber is not available anywhere in the network (this might typically be used by a Presence server for example).

- **603 Decline**   This is similar to the 600 response, but there is no reason code provided.

- **604 Does Not Exist Anymore**   This is sent when the subscriber being addressed is no longer a valid subscriber in the requested network.

- **606 Not Acceptable**   This means that the server is willing to accept the request but is unable to support the session as it is described in the message body. The *WARNING* header provides the reason for not accepting the session.

Global failures typically indicate to the subscriber that the requestor it is attempting to reach is not available anyplace in the network or might not even be a valid subscriber in the network.

Parameters in all of these responses provide additional details as to why a request is being rejected, and may provide other details as well. For exact definitions on how these responses are used, refer to Chapters 5 and 6 on establishing sessions, registering in the IMS, and releasing sessions in progress.

The following section describes all of the headers used in the IMS as defined by the IETF and as defined through extensions by the 3GPP. Most all of these are defined in the RFC 3261 published by the IETF; however, when extensions are defined, they are not found in this RFC. Rather, they are published in various other specifications published by the 3GPP. This makes defining some of these extensions difficult without tracking all of the activities of the 3GPP and the documents published.

SIP Header Fields

Following the start line (or the status line of a response) are various headers providing additional details regarding the request or the response. These headers, then, contain parameters providing the details. This section provides an overview of all of the headers defined (and known at the time of this publication) for use within the IMS. They can be found in RFC 3261, published by the IETF, as well as various specifications published by the 3GPP.

Accept   When an endpoint subscribes to event notification, it uses the *ACCEPT* header to identify what the endpoint can support. For example, if the endpoint is a server building a URI list for a subscriber, then it will send the *SUBSCRIBE* method containing one or more *ACCEPT* headers. Each of the headers will define the format that the server is able to support for the URI list.

The *ACCEPT* header can also be sent within a response to identify media types and format types that are supported by the endpoint. For example, if an *INVITE* is sent with an SDP calling for video in a specific format, but the endpoint does not support video in that format, it can return a response with the *ACCEPT* header identifying the formats supported. The response could be 415 Unsupported Media Type.

The *ACCEPT* header can also be accompanied by the *ACCEPT-ENCODING* header, which identifies the supported encoding for the content.

```
Accept: application/sdp;level=1, application/x-private, text/html
```

Accept Encoding   This header identifies the method of encoding used for the *ACCEPT* header, if encoding is used. It identifies the method by which the receiving endpoint should decode any provided content in place of the SDP.

If the SDP is part of the message, the encoding method is contained within the SDP. However, the *ACCEPT* and *ACCEPT-ENCODING* can be provided as part of a response without using the SDP, identifying what the endpoint is able to support. One example of this usage is within the 415 Unsupported Media Type response.

```
ACCEPT ENCODING: gzip
```

Accept Language   This header is used to indicate when a language other than English is being used in the status codes, SDP, or status responses. It is usually used in conjunction with the *ACCEPT* and *ACCEPT-ENCODING* headers. The default language is always English unless otherwise noted with this header.

```
ACCEPT LANGUAGE: da, en-gb;q=0.8
```

Alert Info   When provided within an *INVITE* message, this header indicates an alternative ringing tone to be used. It can also be used within a `180 Ringing` response to indicate an alternative ringback tone.

```
ALERT INFO: <http://www.ringtones.com/sounds/clapping.wav>
```

Allow   The *ALLOW* header is typically found within an *INVITE* message to indicate what methods are supported by the device sending the *INVITE*. It can also be sent as part of a response to indicate the same for the recipient of an *INVITE*. For example, a response of `405 Method Not Allowed` would contain the *ALLOW* header to indicate what methods are supported by the responding endpoint.

```
ALLOW: INVITE, CANCEL, BYE, ACK, OPTIONS
```

Authorization   When a device sends a request such as an *INVITE* or *REGISTER*, it can include the *AUTHORIZATION* header containing its credentials as part of the authorization process. The credentials are typically stored within the device on a Universal Integrated Circuit Card (UICC).

If the *AUTHORIZATION* header is not present, then the receiving endpoint within the IMS will send a `401 Not Authorized` response. This is called the "challenge" and is sent by registrars and endpoints alike. The response will contain the *WWW-AUTHENTICATE* header describing the authentication scheme that is expected from the device. Registration is explained in detail in Chapter 5.

```
Authorization: Digest username="travis",
realm="raleigh.com",
nonce="dcd98b7102dd2f0e8b11d0f600bfb0c093",
uri="sip:travis@raleigh.com",
qop=auth,
nc=00000001,
cnonce="0a4f113b",
response="6629fae49393a05397450978507c4ef1",
opaque="5ccc069c403ebaf9f0171e9517f40e41"
```

Call-ID   The *Call-ID* provides a unique identifier for all sessions, so proxies (such as Call Session Control Functions) can correlate requests with responses. A single session may require a dialog with multiple entities, in which case the *Call-ID* is used by each of these entities as a reference.

This is not the same as a dialog. Each of the endpoints within a session will have its own dialog (identified by a dialog ID, which is explained in Chapter 5). The *Call-ID* is different than the dialog ID and is used much differently. Think of the *Call-ID* as the identifier for an overall session, while the dialog ID is used by each endpoint to keep track of communications with another endpoint.

```
CALL ID: f81d4fae-15apr-11d0-a765-00a0c91e6bf6@raleigh.com
```

Call-Info   This is a way for SIP to provide additional information about either the called party or the calling party. For example, the calling party may send an ICON as a representation of themselves (as shown in the example that follows), along with a "business card" (using vcard, for example) to the called party. The header field can be sent either in a request (in which case information about the calling party is sent) or in a response to a request (in which case information about the called party is included).

There is potential risk in using this header field. Malicious use of this header field could result in questionable content being sent to a calling or called party. The standards recommend use of this header field only if the source can be authenticated and is a trusted source.

```
CALL INFO: <http://wwww.tekelec.com/travis/photo.jpg>;purpose=icon, <http://
www.tekelec.com/travis/> ;purpose=info
```

Contact   The *CONTACT* header provides additional address information about a subscriber and is used to identify additional addresses that a request can be sent to in the event that the first request fails. When there are multiple addresses identified, the "q" parameter is used to indicate the order in which the addresses should be used. The lowest number indicates the first choice, followed by the next highest, and so on.

The *CONTACT* header allows endpoints to exchange other addresses that may or may not be known by the devices. It also allows redirect servers to indicate alternative addresses being used to route a request to another address.

```
CONTACT: <sip:travis@nc11t2172b.raleigh.com>
```

Content-Disposition   This header identifies how the message body should be interpreted and treated by the receiving endpoint. There are several additional parameters that are part of the *CONTENT-DISPOSITION* header:

- *Session*   Indicates the message body describes a voice call
- *Render*   Indicates the message body should be displayed or rendered to the user
- *Icon*   Indicates that the message body contains an icon that should be displayed
- *Alert*   Indicates the message body contains a ringtone or other similar alerting information (this could be part of the 180 RINGING response, for example)

If this header is missing, and the *CONTENT-TYPE* is *application/sdp,* then it is assumed that the *CONTENT-DISPOSITION* is *session.* Otherwise, it is assumed that the disposition is *render.*

```
CONTENT-DISPOSITION: session
```

Content-Length This header identifies the length of the message body, expressed in octets.

```
CONTENT-LENGTH: 265
```

Content-Transfer-Encoding The *CONTENT-TRANSFER-ENCODING* is used when tunneling a SIP message within the message body of another SIP message. The header identifies the encoding used for the "tunneled" portion of the SIP message.

```
CONTENT-TRANSFER-ENCODING: base64
```

Content-Type This header identifies the type of content contained in the message body. For example, in an *INVITE* for a voice call, the message body will carry the SDP describing the session.

```
CONTENT-TYPE: application/sdp
```

CSeq The *CSeq* header is used for ensuring proper sequencing of transactions during a dialog. The header, which consists of a decimal number followed by the method type, is used by endpoints to allow a device to track the proper response to the proper request.

For example, when an *INVITE* is sent, and a *2xx* response is received, the *ACK* is sent using the same *CSeq* number as the *INVITE,* but with the *ACK* method in place of the *INVITE.* The *CSeq* then identifies requests and responses as different transactions when necessary so that the endpoints can track them separately.

```
CSeq: 3411 INVITE
```

Date The *DATE* header provides the date and time expressed in GMT. This information can be used by systems that do not have battery backup, for example, as long as they are able to determine their GMT offset. For example, a portable gaming device that gets shut off and loses its date and time can be reset when it is powered back up and registers with the network.

This can also be used to provide network date and time to end devices such as cell phones and PDAs.

```
DATE: Mon, 10 Aug 2007 14:33:03 GMT
```

Error-Info This header is used to provide additional information regarding an error response and is used by the endpoint to provide notification to the user. For example, the error code itself could be displayed on the user's device, while the header provides the location of a recording to be played back to the user.

This provides more flexibility to the operator in terms of how to implement user notification and service tones/recordings.

```
SIP/2.0 404 Not Found
Error-Info: <sip:out-of-service-recording@raleigh.com
```

Event
Used with the *NOTIFY* method, the *EVENT* header identifies the event that caused a change in registration status. For example, if a Presence server has subscribed to event notification, when a subscriber device changes its registration (moves to another location resulting in a new IP address being assigned) the *NOTIFY* method sends the new registration information. The *EVENT* header then provides the reason (*reg,* for registration) the registration was changed.

```
EVENT: reg
```

Expires   The *EXPIRES* header is used for multiple purposes. It allows the sender to express when an event (such as a user's registration) is to expire. For example, when the *EXPIRES* header is sent within a *REGISTER* message, the *EXPIRES* value indicates when the registration is to expire. The user must then register again prior to the *EXPIRES* value.

The use of this header varies upon its implementation. It is flexible in that it can be used to express any expiration (not just registration) based on the method or request it is contained within. The value is expressed in decimal as seconds.

```
EXPIRES: 3
```

From   The *FROM* header identifies the initiator of a request. There are a number of ways it can be expressed, including a display name or the URI. The URI can be a TEL URI as well. This field is primarily for user consumption and is not used by the network for routing. For example, a device could use this information for name display. However, care should be taken when using this header, as it can easily be compromised.

When the URI is being presented along with the name display, the URI is contained within brackets, and the name display is contained within quotations.

```
FROM: "Travis Russell" <sip:travisruss@aol.com>
```

In-Reply-To   This header is used for applications where calls that previously were unanswered are returned. The content of this header is the *CALL-ID* of the call being returned. It could be used by call distributors and other applications where users want control over the calls that are accepted and returned. It can also be used by voicemail systems and other, similar applications where the user misses a call and wants to return the missed call.

```
IN-REPLY-TO: 89222@aol.com
```

Max-Forwards   *MAX-FORWARDS* is used to prevent looping of messages. When a message is created (either a request or a response), this header is set to 70. As the message passes through each proxy in the network, the value is decremented by one, until the value reaches 0. If a proxy receives a message with a *MAX-FORWARDS* value of 0, the message is discarded.

This can also be used by systems that are measuring or monitoring the network to troubleshoot routing of SIP messages. By measuring the value of the *MAX-FORWARDS* header at various points in the network, one could derive where looping is occurring and identify proxies that are causing circular routing.

```
MAX-FORWARDS: 70
```

Min-Expires   This header identifies the minimum time allowed for the expiration of a value such as *CONTACT* within a proxy, such as a registrar (the S-CSCF in the IMS). The value is a decimal number expressed in seconds.

```
MIN-EXPIRES: 60
```

MIME Version   This header identifies the version of MIME used within the message body if there is content provided in MIME format. This allows the receiving device to properly render or decode the content as received.

```
MIME-VERSION: 1.0
```

Organization   This can be used by the operator within the IMS to identify the name of the operator and domain (Jack's Telephone Company, e.g.). It can also be used by users to identify the name of their company for display purposes. It identifies the name of the organization that initiated the request (or response) and is used for informational purposes (such as display on a device).

```
ORGANIZATION: Jack's Telephone Company
```

P-Access-Network-Info   This header is used to identify the access method used to access the IMS network. This is an extension of the SIP protocol as submitted by the 3GPP and identifies the type of network (such as GPRS) that was used to gain access into the IMS. In most SIP networks, this information is not needed and, in fact, not of any interest. It was added specifically for IMS networks where access is provided by means other than IP.

As such, this information should be treated with extreme care, as it could be considered as very sensitive information. It could be used by competitors to determine how many times subscribers are using various access technologies, for example, and collected to gain sensitive subscriber behavioral data.

For this reason the 3GPP recommendations suggest that this header not be shared between networks but only be used within the home network. This means that prior to forwarding a request to another network, the CSCF must delete the header from the request.

It can also be used for billing purposes where special tariffs or rates may apply, depending on the type of access used by the user. For example, the operator may have a charge for Wi-Fi that is separate from the charge for GPRS. Of course, the billing systems can then parse this information and apply it in any fashion they like (such as by charging according to the type of session and technology used, and the session duration).

```
P-ACCESS-NETWORK-INFO: gprs
```

P-Asserted-Identity This header is inserted after authentication has been complete. The URI of the authenticated subscriber is provided in the header as a trusted identity. The receiving entities can be assured that the identity contained in the *P-ASSERTED-IDENTITY* header can be trusted and that this is the true identity of the user.

```
P-ASSERTED-IDENTITY: travisruss@aol.com
```

P-Associated-URI Upon registration, the S-CSCF in the IMS (or the SIP registrar in a VoIP network) can use this header to return any URIs that are associated with the registration. The association is assigned by the operator, based on the subscription. For example, a user may have several identities within one operator network based on multiple uses. There may be one URI and identity used for personal use, while another is used for work.

The registrar retrieves these associated URIs from the HSS during the registration process and sends them to the device that is registering so that it may save in its cache the other URIs it can answer. A complete description of this process is found in Chapter 5.

```
P-ASSOCIATED-URI: travisruss@aol.com; russell@tekelec.com
```

P-Called-Party-ID When an *INVITE* is forwarded to another address for a registered subscriber, it is impossible for the user device to determine for what registration (or what identity—personal, business, etc.) the original session was intended. This is because the original *INVITE* contains the called party within the request-URI, and when the session is forwarded by a proxy, the request-URI is changed to reflect the new destination.

For this reason the *P-CALLED-PARTY-ID* header was added as an extension to the SIP protocol. This allows the proxies to insert the original identity that was "dialed" or called, allowing the receiving device to treat the call accordingly.

For example, if the *P-CALLED-PARTY-ID* reflects that the original session request was sent to the user's business URI or identity but is being received on his or her personal identity, then the device could apply special ringing or send the call directly to voicemail, based on call treatment defined by the user.

```
P-CALLED-PARTY-ID: sip:russell@tekelec.com
```

P-Charging-Function-Address This is another header that was added by the 3GPP specifically for use within the IMS. In fact this header is applicable to IMS networks using the 3GPP charging architecture. Within this architecture there are many entities that have the ability to generate charging records for the purpose of billing the subscriber for services rendered. These entities include the various access points (SGSN/GGSN) and the CSFCs within the network.

The billing data is collected by one of two functions within the IMS charging architecture: the event collection function (ECF) and the charging collection function (CCF). These functions may be distributed throughout the network in a manner dependent on the size of the network and the specific implementation. They are identified by their IP addresses, and each of the network entities responsible for generating the billing data for the charging functions is homed to one of the collection functions.

The use of this header is generally restricted to within the home network, or between two trusted domains. Since it contains the addresses of the entities responsible for the collection of billing data, it would not be in the best interest of the operator to allow this information outside of its network. Therefore, the I-CSCF is responsible for deleting the header prior to forwarding to another network.

The various entities that generate the billing data do so for sake of either postpaid billing or prepaid billing. This determines where the data is sent. There are also considerations for redundancy and distributive deployment, which means collection functions could be duplicated in various parts of the network. This means then, that the entities providing the billing data must be aware of the charging entities supporting their part of the network as identified by their IP addresses.

This header then provides the addresses of those entities that billing data should be sent. The use of this information is detailed in Chapter 7.

```
P-CHARGING-FUNCTION-ADDRESS: ccf=192.1.5.34; ccf=191.6.4.98; ecf=192.9.3.3;
ecf=192.1.9.3
```

P-Charging-Vector   This header is also used for IMS charging. It is inserted by a SIP proxy (such as a CSCF) when a request or response is received. It is strictly up to local policy as to how and when the header is inserted, but typically within an IMS architecture this would be inserted by the gateways and each of the CSCF entities throughout the network.

The information provided in this header is used to identify where a charging record has been created (as identified by the address in the *icid-generated-at=* parameter) and the network responsible for generating the record. The originating network is identified in the *orig-ioi=* parameter.

There is also an identifier assigned for correlation purposes. This allows charging entities to correlate this with other charging records for the same sessions. It is a globally unique identifier provided in the *icid-value=* parameter.

Normally this header would only be provided within a trusted domain. That is, this would only be shared within the home network, and not with other, external networks. However, there are provisions that would allow this information to be passed to another network through the I-CSCF provided local policy permits. Typically the two operators would have to have a trusted relationship (and an intercarrier billing arrangement) before this would be allowed.

```
P-CHARGING-VECTOR: icid-value=9876a4321ce; icid-generated-at=193.1.0.5;
orig-ioi=aol.net
```

P-Preferred-Identity   This header can be provided by a device to indicate the identity that is preferred for the specified subscription, based on the user's interaction with a Presence server or other application. For example, a subscriber may wish that all business calls be routed to this address rather than the URI provided in the request-URI.

```
P-PREFERRED-IDENTITY: russell@tekelec.com
```

P-Visited-Network-ID   The *P-VISITED-NETWORK-ID* allows networks outside of a subscriber's home network to identify themselves to a home network when transiting requests and responses for a roaming user. This function is somewhat supported today in the SS7 ISUP protocol using the Transit Carried ID, although its implementation has been sparse at best.

This parameter provides operators some alternatives. For example, the problem today with intercarrier compensation in wireline networks is in identifying the first and second transit networks prior to a call reaching the destination network. There is no means of identifying the first transiting network, and if the call transits yet another network, the destination (or terminating) network is left no choice but to trust and charge the adjacent carrier for termination.

This header would allow each transiting carrier to place its identity prior to forwarding the request on to the next network. The home network would then be able to determine which networks the session request (and ultimately the call itself) transited, and would be able to then apply the appropriate intercarrier fees.

Use of this header is dependent on intercarrier agreements between the various carriers, since first, the network identifier must be known to the home network, and second, the transiting carrier has no obligation to provide this information. It is strictly voluntary as an optional protocol header.

The network identifier must also be globally unique. That requires some third-party administrator to prevent duplication of identifiers, much like what is implemented for the Domain Name Service (DNS).

The *VIA* header could also provide the same information, but it may be undesirable because it may require queries to the DNS when the operator inserts its IP address. This would require a reverse DNS to determine the domain, which would then, of course, identify the operator's network. As an alternative, this header would be used in place of the *VIA* header (the receiving proxy would simply ignore the *VIA* header and use the *P-VISITED-NETWORK-ID* to determine which networks were transited).

```
P-VISITED-NETWORK-ID: joesnetwork.com
```

Privacy   This header is used to identify when a subscriber wishes to hide his or her identity from other networks. This could be the case when a subscriber wishes to maintain secrecy for certain communications, but of course it can be used for illegitimate purposes as well.

The header is provided in the specifications for legitimate purposes, allowing subscribers to provide a different identity in the *request-URI* than they do in the *TO* header.

The *TO* header, of course, is displayed to the calling party, while the *request-URI* is used by the network for routing and authentication purposes.

```
PRIVACY: ID
```

Priority This header was added for providing emergency calls priority over normal traffic; however, the implementation of this header is operator specific. Therefore, since it has not been defined for global usage, it may or may not provide priority service for, say, emergency calls (such as 911 calls in the U.S.).

Other work is being done to define a mechanism for preemption and emergency priority services in the event of a catastrophe that are not defined here (this is still a work in progress), but this header is explained so that operators can provide an on-net-type service if they wish.

For example, an operator may elect to provide an Emergency Alert System within its own network as an additional service to its subscribers. To ensure its emergency messages were routed as a high priority, it could implement the use of the *PRIORITY* header. This would allow an operator to give a higher priority to its emergency messages than it would to normal messages.

The RFC defines some suggested values for the *PRIORITY* header of *"normal," "non-urgent," "urgent,"* and *"emergency."* Their use is completely dependent on the operator, although the IETF does recommend that the use of "emergency" be limited to life-threatening situations.

```
SUBJECT: Thunderstorm Warning for Johnston County
PRIORITY: Urgent
```

Proxy-Authenticate The *PROXY-AUTHENTICATE* header is provided as a challenge by a proxy (such as an application server or a CSCF). When a request is sent by the device, the proxy can send a 407 Proxy Authorization Required response, containing this header. The device then must return a response containing the *PROXY-AUTHORIZATION* header, with the proper credentials.

This could be used in cases where the operator wishes to authenticate a user every time that user accesses its call control server, to prevent unauthorized access to a server used to define call treatment and voicemail services by the subscriber.

```
PROXY-AUTHENTICATE: Digest Realm="tekelec.com,"
domain= "sip:Verizon.com", qop="auth",
nonce= "a73dd646abc898763fbade98129e763547",
opaque= "", stale= FALSE, algorithm= MD5
```

Proxy-Authorization The *PROXY-AUTHORIZATION* header is used when a device is authenticating itself with the challenging proxy in response to a 407 Proxy Authentication Required. It is within this header that the user provides his or her credentials, known only to the device and the network provider.

None of the data provided for authentication is human-consumable data. This is data that is created using the various algorithms and keys that are provided by the network provider at subscription time and embedded within the device (usually within the UICC or SIM card).

This process is different than registration, where the network registrar performs the authentication. This procedure and its associated headers provide the operator with an added measure of security to prevent unauthorized access to network services, even after registration.

The use of this header is implementation specific and is not required, but it does add an additional level of security for the operator and an additional layer of protection to select services housed on application servers.

```
PROXY-AUTHORIZATION:Digest username= "Travis", realm= "Verizon.com",
Nonce= "c60f3082ee1212b402a21831ae"
Response= "43536ff4355tcc45567"
```

Proxy-Require This is similar to the *REQUIRE* header, except that it is used by proxies to communicate the extensions and capabilities that must be supported by the device. The proxy sends this header in response to requests to communicate what it requires of the user device when supporting a session.

It is different from the *REQUIRE* header, which is used between a client and a server (non-proxies). The client and the server could be resident on two devices, or they could be resident on application servers.

```
PROXY-REQUIRE:foo
```

Record-Route The *RECORD-ROUTE* header is used for strict routing within the IMS. As a request is routed through the network, each proxy (CSCF or other entities such as MGCFs) inserts this header along with its address into the request. When the request is received by the destination, it uses the *RECORD-ROUTE* to send a response.

Think of *RECORD-ROUTE* as the header used to create a routing list for a subscriber device. One use for this in the IMS prevents hijacking of sessions when routing is enforced. When a device registers with the network, *RECORD-ROUTE* is used as the *REGISTER* is routed through the network, and the various entities used to route the *REGISTER* to the S-CSCF enter their addresses prior to forwarding to the next entity.

When the S-CSCF receives the *REGISTER,* it then uses the *RECORD-ROUTE* headers to create a route list for the user. All responses are then sent using the same route as recorded. This route is stored as part of the registration, so that all subsequent requests and responses use the same route.

This form of strict routing ensures that a man-in-the-middle attack cannot be used to hijack a subscriber's registration, for example. It ensures that all requests and responses are sent through the same path used for the registration to reach the user.

The P-CSCF then uses this route list to enforce routing messages to the user. See Chapter 7 for more details on this form of routing and the security procedures that use this header.

Reply-To The *REPLY-TO* header is inserted by a user device upon receipt of a request. It is used to communicate the direct address of the device for all subsequent responses and requests throughout a dialog. This in essence would then allow responses to bypass the various proxies within the network and allow routing directly to the device.

Within an IMS domain, there may be concerns about routing directly to a device, bypassing the CSCF within the network. In fact, this form of loose routing is not defined within the IMS. The IMS procedures call for strict routing to ensure that all requests and responses always follow the same path used during registration.

The *REPLY-TO* header is still supported within the IMS, but it is not necessarily used to route responses. It is simply used to identify the direct address of the device, but requests and responses are still routed through the CSCF entities within the network. Other proxies may be bypassed, however.

```
REPLY-TO: Travis Russell <sip:travisruss@aol.com>
```

Require   This header is used by entities to identify any SIP extensions that must be supported by the other entities. This is similar to the *PROXY-REQUIRE* header discussed previously, with the exception that this is sent by devices rather than proxies.

This header is needed because SIP extensions may not be supported in every network (and consequently by every device). If a device is sending a request using specific services that are enabled through a SIP extension, then those extensions must be identified to the other endpoint to ensure the session can be supported.

```
REQUIRE: 100 REL
```

Retry-After   This header is used with the 500 Server Internal Error and 503 Service Unavailable responses to indicate to the requestor the duration that the server is expected to be unavailable. It can also be used with 4*xx* responses to indicate when the called party expects to be available again.

The parameters would indicate the duration of time and possibly the time when the calling party could try to reach the called party again. Optionally a message such as "In a Meeting" can be provided as part of the response to identify why the party is rejecting a call or why the called party is not available.

```
RETRY-AFTER: 19000; duration=3600
```

Route   This header is used along with the *RECORD-ROUTE* header when strict routing is implemented (as is the case in the IMS). When a request is being sent, the *RECORD-ROUTE* header records the address of each of the entities in the call path. The response then inserts these addresses in the *ROUTE* headers (there are typically multiple headers).

The headers are listed in the order of the route. In other words, the addresses are shown in the same order they are routed through. The routers then use this for routing the responses to the next hop in the network.

```
ROUTE: <sip:Raleigh@bellhead.com>
```

Server   This header allows the server (the recipient of a request) to communicate the software version being used by the server to process the request. There are many uses for this information, depending on implementation.

Care should be taken with this header, as information regarding software versions could be a security risk. Hackers could use this information to obtain the version of the Symbian operating system resident on a cell phone, for example, and then send a virus or Trojan to that device. Operators should use encryption to prevent this information from being read by any other than the end devices in a session.

```
SERVER: Symbian OS 8.0
```

Subject   This is much like the subject line in e-mail. It is provided as a means of sharing additional information about the session for display to the user. For example, if used for an emergency broadcast feature, the *SUBJECT* header would contain the actual alert message (such as "Thunderstorm warnings for Johnston County") while the *PRIORITY* header would contain the priority for the session.

The *SUBJECT* header can also contain text displayed on the end device for a call. The sender may wish to have the message "Answer the phone!" pop up when the call begins ringing on the called party's phone (subject to call servers supporting such a function). There are many other implementations for the *SUBJECT* header, depending on operator implementation.

```
SUBJECT: Anyone Home?
```

Supported   This header is used by either endpoint to communicate the SIP extensions and capabilities supported by the sender. This is different than *REQUIRED,* which is sent to communicate the SIP extensions that must be supported for a session.

The *SUPPORTED* header would be sent in response to *REQUIRED* or *PROXY-REQUIRE* to communicate the extensions supported, or it would be included in an unsolicited request/response.

```
SUPPORTED: 100 REL
```

Timestamp   While the use of this header is not fully defined in RFC 3261, it does allow for the proxy or any other entity to enter a timestamp, which could then be used for determining round-trip time (RTT). This would require each proxy to enter the timestamp as the request/response was forwarded to the next hop, and the endpoint to have the ability to process this information.

```
TIMESTAMP: 50
```

To   The *TO* header serves as an address field for human consumption only. This means that the information is not used by the network for routing purposes. It is only there for display on end devices.

This information is also easily spoofed and therefore should never be trusted. The information can be populated by the originating requestor, or it can be modified by proxies along the call path. This header is analogous with the *TO* header in an e-mail.

The first part of this header contains the display name to appear in the receiving device's display. The portion contained within "<>" is the actual address of the recipient. The "*tag*" is used as part of the dialog ID and is used by the receiving device to correlate this request with other requests/responses.

```
TO: Travis Russell<sip:russell@tekelec.com>;tag=1df789jkf
```

Unsupported Similar to the *SUPPORTED* header, this header is used to identify extensions that are not supported by an entity.

```
UNSUPPORTED: 100 rel
```

User-Agent This header is like the *SERVER* header, providing the software version of the SIP user agent processing a request. The two headers can be used together to identify the software versions on both devices. This information can then be used by operators for profiling and other applications.

For example, in today's networks cell phones send an International Mobile Equipment Identifier (IMEI) identifying the make and model of the device, and configuration information about the device. Operators that collect this information can then use the data to track the activities of the subscriber by reference to their phone make and model. They can use these statistics to better understand the behaviors of the subscribers in terms of their phone models.

They can also determine how the various phones are being used, and whether or not subscribers are using all capabilities of the device. If a device identified supports video, the operator can track the actual usage of all video-enabled devices to determine how many actually use this capability, when they use the capability, and so on.

This data could even be used for promotional campaigns to alert all users of certain models of devices of new features and applications that are being offered for their devices. Application servers can also process this information and use it for various applications.

```
USER-AGENT: Vista Beta1.5
```

Via The *VIA* header is a means of recording the path that a request takes to reach its destination so that all responses follow the same path. When using loose routing, the *VIA* header ensures that responses are received by stateful proxies in the call path. In strict routing, this is used in conjunction with the *RECORD-ROUTE* header.

The difference is that the *VIA* header is used by the proxies to determine the next hop in the network for a response, while *RECORD-ROUTE* is actually used to create a route list that will be used for routing all requests and responses to a device throughout the life of its registration. The *VIA* header is never stored as part of the registration but is only used for by proxies for routing.

```
VIA: SIP/2.0/UDP pchome101@aol.com:5060; branch= z9hG4bK713a2
```

Warning   Warnings indicate problems in processing the session description itself. They are different than error responses, since the session itself is being processed. The *WARNING* header contains a text description identifying the purpose of the header as defined here:

- 300 Incompatible network protocol
- 301 Incompatible network address formats
- 302 Incompatible transport protocol
- 303 Incompatible bandwidth units
- 304 Media type not available
- 305 Incompatible media format
- 306 Attribute not understood
- 307 Session description parameter not understood
- 330 Multicast not available
- 331 Unicast not available
- 370 Insufficient bandwidth
- 399 Miscellaneous warning

The definition of these warnings is outside the scope of this book. Their use is defined in RFC 3261. The preceding text values are suggested and not mandatory. Operators can define their own text values for these warnings for their own implementations.

```
WARNING: 307 tekelec.com "Session parameter 'foo' not understood"
```

WWW-Authenticate   This header is used as a challenge to an entity sending a request. The challenge is carried in the response to a request with this header. The response will contain the AUTHENTICATE header containing the proper credentials.

```
Authorization: Digest realm="raleigh.com",
domain= "sip:tekelec.com", qop="auth",
nonce="dcd98b7102dd2f0e8b11d0f600bfb0c093",
opaque="5ccc069c403ebaf9f0171e9517f40e41"
```

Session Description Protocol (SDP)

The session description protocol is what describes the session being requested. For example, for a voice call the codecs that are used to convert the voice transmission to digital must be identified in the SDP so that the receiving end knows how to decode the voice at the other end.

The SDP is carried in the message body of a SIP request/response. Each attribute line consists of an attribute identified by a single letter, followed by a value. The attributes defined for SDP follow. They are defined by these categories:

- Session Level Description

- Time Description
- Media Description

Session Level Description

- **v = protocol version** This is the version of SDP being used to create the SDP.
- **o = owner/creator and session identifier** The identity of the session initiator.
- **s = session name** Optional name that can be given to the session, such as "Travis's Webinar."
- **i = session information** Additional information that the creator of the session wishes to share with participants
- **u = URI of description** This contains the URI of a Web site that may contain additional information about the session. For example, if the session is a Webinar, the Web site is where participants can go for additional information about the Webinar (such as login information).
- **e = e-mail address** This usually contains the e-mail address of the creator, where participants can acquire more information.
- **p = phone number** The contact phone number where more information can be provided about the session.
- **c = connection information** Additional information about the connection for the session.
- **b = bandwidth information** The amount of bandwidth to be provided for the session.
- **z = time zone adjustments** Any time zone adjustments to be considered.
- **k = encryption keys** The encryption keys for the session.
- **a = zero or more session attribute lines** The number of attribute lines in the SDP.

Time Description

- **t = time the session is active** What time does the session start, for example.
- **r = zero or more repeat times** How many times the session repeats.

Media Description

- **m = media name and transport address** Name of the media, if applicable.
- **i = media title** Title of the media (Travis's Vacation Film, e.g.).

- **c = connection information**   Additional connection information.
- **b = bandwidth information**   Bandwidth required to support the media.
- **k = encryption key**   Encryption keys required.
- **a = zero or more attribute lines**   Number of attribute lines provided.

The attribute lines themselves provide additional information about the session. Some are self-explanatory. They are provided here simply for reference. Their usage is explained in another text, *Session Initiation Protocol (SIP): Controlling Convergent Networks* (McGraw-Hill, forthcoming).

Media Attribute Lines (a=)

- cat = category
- keywds = keywords
- tool = name and version of tool
- ptime = packet time
- maxptime = maximum packet time
- recvonly = receive-only mode
- sendrecv = send and receive mode
- orient = whiteboard orientation
- type = conference type
- charset = character set
- sdplang = language tag
- lang = language tag
- framerate = frame rate
- quality = quality
- fmtp = format-specific parameters
- rtpmap = rtpmap attribute
- curr = current status attribute
- des = desired-status attribute
- conf = confirm-status attribute
- mid = media stream identification attribute
- group = group attribute

The SDP, then, is used to describe the parameters for a session, whether it is a voice call or a Webinar. If a conference call provides Webinar-type service, there are many attributes that must be defined, but keep in mind that each part of a Webinar is a separate session.

In other words, the voice portion of the Webinar is one session, while the display of a whiteboard application is a different session. Likewise, if a document is shared during the Webinar, yet another session is established. Each session then would carry its own SDP describing that portion of the session.

This means that for a complex session such as a conference call or Webinar, one could not see every aspect of the session by capturing just one SDP. It would require capturing all of the *INVITES* for each portion of the session and correlating these all together.

The SDP for each of the sessions would then be correlated to provide a complete view of the overall session. This becomes a complex task in itself for troubleshooting, since technicians will need to be able to determine what session they need to troubleshoot. Simply tracing by the end-user identity will produce many sessions in progress, each session being used for different purposes.

The SDP could be used to determine what each session is for, and the type of each of the sessions. This should be a capability provided in the OSS/BSS. From there the technician could troubleshoot each individual session to determine if there was an error with a part of a session or with the overall call.

Remember that the SDP is not required for all sessions. It is required only for sessions where two endpoints are going to enter into a dialog with one another. A simple example of this is a voice call.

If there is no SDP in the message body of a SIP request/response, other content could be present. For example, SIP can be used to carry the content for a text message (using the *MESSAGE* method). The absence of an SDP does not mean that there is no session content.

4

Addressing in the IMS

The telephone number is dead. This may sound strange when we still dial these numbers every day, but look at your business card. While there are still several telephone numbers on your business card, they are followed by your e-mail address, as well as the URL of your company's Web site. Wouldn't life be easier with just one identity?

The issue is our networks don't support all forms of communications. There is the telephone network that requires telephone numbers to route calls. There is the Internet that requires e-mail addresses to route mail. The Web requires Universal Resource Locators (URLs) to identify Web sites.

If there could be one network, supporting all connections, there would only be one identity needed for all communications. We are now getting closer to being able to realize this. The IMS model allows operators to use one account to support everything that a subscriber needs to communicate.

Not too long ago an e-mail address was unusual to find on business cards. Not everyone had one. Today, most people I know have several addresses. They use one (or more) for personal use, they have one for business use, and of course there is the family e-mail address. We dole these addresses out to our friends, family, and business associates depending on the need.

But as voice services become part of the same service that delivers us our e-mail, it will quickly begin to make telephone numbers obsolete. After all, if I could have one service that took care of everything I use for communications, why have different forms of identity for each one? Why use an impersonal number when I could use the more personal identifier found in our e-mail addresses?

Also consider that the very technologies designed to deliver our e-mail are now being adapted to deliver our voice, video, and messaging. The very systems that form these networks operate differently than the telephone switches that rely on telephone numbers.

To better understand this shift in identities, it may help to better understand where telephone numbers came from and how they were used in the earliest days of telephony. In days far gone, when one wanted to connect to a neighbor or family member using the telephone, all they needed was to pick up the phone, wait for an operator, and give the name of the person they wanted to connect with.

But as communities grew, and the cord boards used to connect calls were replaced with switches, numbers began to replace our names. The numbers were used to drive electromechanical relays that literally connected the wires from one home to another across the entire network. These relays made up an intricate maze of circuits and connections, all based on the dialed digits.

The digits themselves began taking on meaning. Communities were once referred to as "exchanges", serviced by a switch. The switch was identified by the first set of digits dialed. Once the switch was reached, the remaining digits identified the subscriber to be reached within that switch.

This allowed telephone companies to interconnect hundreds of switches to one another, automating the entire connection process. Operators were no longer needed, and subscribers could dial any place they wanted (although in the beginning they could only make local calls).

As the technology evolved and switches became digital, there was much more that could be done with telephone numbers. The numbers were extended in length to support long-distance and international dialing, further removing the human factor from the telephone experience.

Believe it or not, telephone numbers have long been viewed as unfriendly and difficult to remember. Consumers have grown so accustomed to this form of identity, however, that they soon began to memorize the exchanges by the telephone numbers.

For example, consumers know that the 212 area code is New York City. When a call is received bearing the caller ID of 212, the call recipient knows that the caller is in New York. When number portability was first introduced in the U.S., many fought against geographical portability because they believed that consumers would reject the idea because they would no longer be able to tell the origin of a call.

How ironic that in the world of the Internet, the creators sought to maintain the personal touch. They recognized that memorizing yet another set of numbers would not work for the Internet. Using IP addresses for the routing of e-mail would be far too much for consumers and therefore they would not use the service.

Certainly the Internet (and all its underlying technology) ultimately relies on a set of numbers for addressing. For this reason another form of identity was developed, and the function of resolving these identities to an IP address left to the network.

Hence was born the present form of the e-mail address, or the Universal Resource Identifier (URI). The subscriber, of course, has complete control over their e-mail addresses, but the portion behind the "@" symbol is reserved for use by the service provider. Not only can you identify the subscriber, you can also identify the service provider providing services to that subscriber.

The URI still has to be changed to a number format for use within the Internet, and this is true for the IMS as well. We won't go into the specifics of how that is done here, as this is discussed in more detail in later chapters. However, we will say that the URI is resolved into an IP address using a service called the Domain Name System (DNS).

The concept of identifying subscribers extends to their devices as well. It is important for telecommunications service providers to be able to validate that the device accessing their network is authorized, and the subscription is current and legitimate. This is to prevent service theft.

But identifying devices also helps deliver unique services based on the device type. For example, a subscriber may wish for e-mail to be sent to his or her Blackberry, but for voice calls to be sent to a separate cell phone. The subscriber may also wish to have personal calls sent to a different cell phone.

This need alone drives the requirements for multiple identities within the IMS model as we will discuss in this chapter. These identities also provide a lot of flexibility for the subscriber. For example, a subscriber can have one identifier for personal use, while maintaining another identity for business use. All of their identities are supported by the same service provider, who then manages the routing of calls and e-mails to the various identities based on their profiles.

All of these are supported through one subscription, rather than forcing a subscriber to have multiple subscriptions for different needs. This is an important concept to understand and something that will become more clear as we discuss how identities are defined and used within the network.

Device Identity in the IMS

There are different levels of addressing used within any network. The lowest form of addressing in a network is the actual address of the connection itself. The connection address is assigned, usually, by the network and is used to route sessions or calls to the device in the network. Depending on the access method used, the connection address may represent a location for the device. The connection address format will also depend on the method of access.

The wireline telephone network uses telephone numbers assigned to subscribers and their devices with a connection within a switch, referred to as the subscriber line. A subscriber line is a physical connection within the switch serving the connection. It is static, does not change, and sits idle until the subscriber on the other end wishes to use the connection (or until someone tries to call the subscriber). There is no other public identity for the subscriber, and as long as the subscriber does not move to another place, this methodology works fine.

However, if subscribers are not home, there is only one way to reach them. You have to call them using their work number, which is probably provided by another company. If they are not at their office, you then have to call their cell phone number, which is also provided by another company.

Each of these companies will have an identity for the subscriber that only they know, for billing purposes. So in reality, a subscriber has multiple identities today for their device. The identity is not necessarily for the subscriber, but the actual device used to reach the subscriber.

Imagine now if you could have one number assigned to you as a subscriber, and calls would find you no matter where you were. The telephone number is now associated to you the subscriber, rather than the device. All anyone would need to know is your identity, and the network would take care of routing to the right device based on your location. This means that the present form of identification must change.

There are several parts to a telephone number, just as there are many different parts to an IP address. The telephone number ends with the subscriber line identifier, preceded by the address of the serving switch (or exchange as they were once referred to). This is preceded by a city code, and a country code. The switches in the network dissect the telephone number in sections, and use these various portions of the address to determine how to route the call (based on routing tables that identify the circuit[s] to use to connect to the next hop in the call).

By far this is the simplest form of addressing, as the numbers remain static and generally remain with a subscription for a long time. I am sure you are well accustomed to telephone numbers and how they are used within the switched network, so we won't go into a lot of detail here.

One thing I will mention here about telephone numbers concerns security and authentication. In a fixed-line network, authentication takes place when the subscriber orders telephone service. There is no need, therefore, for the network to know if the subscriber is legitimate; the line is associated with a subscription and a fixed physical address. The wiring used to connect to the subscriber address further ensures that the subscription need not be authenticated. The wiring always goes to the same address used when the service was ordered.

This is not the case with IP. A subscriber may be using his or her laptop to access the network from anyplace; therefore, it is impossible to ascertain if the user accessing the network is the same user associated with the subscription being used. Authentication is a challenge in these networks.

The physical address is dynamic, assigned by the access network when the subscriber connects to the Internet (or to the local IP network). This means that the identity of the subscriber cannot be authenticated without additional mechanisms where credentials can be exchanged between the network and the subscriber device. This is exactly what happens in the wireless world, but not yet in the Internet.

In a wireless network, there are several addresses that are assigned to the device and the subscription that are communicated to the wireless network. The phone is identified in the CDMA network via the Equipment Serial Number (ESN), which is embedded within firmware of the phone itself. This number identifies the type (make and model) of phone, as well as a unique serial number for the phone. This is used to verify that the device is not stolen.

The subscription in the CDMA network is identified through a Mobile Identification Number (MIN); this is usually derived from the phone number, although it does not have to be. The MIN is used by the network to identify the subscription, but this does not apply to the IMS. The MIN would only be used to identify the subscription within the access network and not the IMS.

In the GSM network, a similar mechanism is used. The phone is assigned an International Mobile Equipment Identity (IMEI) to identify the make and model of the GSM phone. But its use is much different than the ESN in CDMA networks. In the GSM network, the IMEI only identifies the phone and is not tied to the subscription. The subscription is identified through another identity, known as the International Mobile Subscriber Identity (IMSI).

The IMSI is programmed into the Subscriber Identity Module (SIM), which is found inside the GSM phone. SIM cards can be removed and used in different phones, which is why there is no link between the IMEI and the subscriber. This allows subscribers within a GSM network to move from device to device but maintain their identity within the GSM network. This is an important concept because it is where the IMS concept of identity and mobility was born.

Credentials stored in the SIM card are used to authenticate the subscriber. These credentials include a cipher key for encryption across-the-air interface. One of the major concerns of wireless operators is eavesdropping on the air interface. This is easy to accomplish with a device programmed to intercept wireless transmissions. It is for this reason that the transmissions over the air interface are encrypted.

An IMSI is not for public consumption and is sent as infrequently as possible. Usually the IMSI is sent during registration with the network, but if a subscriber is roaming, this number may be replaced by a Temporary Mobile Subscriber Identity (TMSI). We won't go into detail here as to when that is the case.

The IMSI is comprised of a Mobile Country Code (MCC), which is usually three digits long, followed by two or three digits identifying the Mobile Network Code (MNC), and finally the unique Mobile Subscriber Identity Number (MSIN). The Mobile Subscriber International Services Digital Network (MSISDN) is the telephone number that gets assigned to the subscriber and becomes the subscription identity for the subscription and the public address, while the IMSI remains known only to the network.

So we can see that even within the wireless network there are multiple identities. One identity known only to the network provider and hidden from competitive providers; another is for public consumption. But still these identities are for voice services only and do not include e-mail and other services.

Again, as with the MIN in the CDMA network, the IMSI and the MSISDN are not used within the IMS domain for identity. The IMS uses its own identity for subscriptions, but if CDMA or GSM networks are used to access the IMS, these identities may end up embedded in the IMS identities in some form.

For example, a subscriber URI may be used to access e-mail, support voice calls, and provide messaging services. This one identity provides the subscriber with one identity instead of many, and through the IMS the subscriber may receive multiple services through one provider.

The IP domain is another story. A data device such as a computer or PDA uses IP addressing as an access identity. The IP address is assigned by the network to which the device connects. Unlike telephone numbers in the switched network, IP addresses are typically allocated on a connection-by-connection basis. This allows operators to maintain enough IP addresses for the connections they support; instead of enough IP addresses for each subscriber to have his or her own unique address (dynamic addressing is not always used, as some types of accounts may use static IP addressing).

Device addressing is supported through protocols such as Ethernet (the MAC address is the device address), but these are not used within the IP domain because the business model is very different. In IP networks, the service provider does not care about the device accessing their network or the identity of the circuit being used to access their services.

In the IP domain, such as the Internet, revenues are generated through the purchase of videos, music, and other goods. Advertising on the various commercial sites provides yet another revenue stream. So the subscriber identity does not include the device identity in these networks.

Static addressing is pretty straightforward. The service provider identifies an available IP address, which then must be configured in the software of the device. It will identify itself using the assigned IP address each time that device is powered on.

Static addressing puts a strain on network providers with many subscribers, because there is a limit to the number of addresses that can be supported. Not to mention that the address becomes attached to a specific location (determined by the wiring of the network).

Dynamic addressing is assigned through a Dynamic Addressing Host using the Dynamic Host Configuration Protocol (DHCP). DHCP is a protocol that is used to exchange IP addresses with configuration servers hosted in the network. This allows operators to re-use IP addresses, rather than use dedicated IP addresses for each subscriber. Dynamic addressing is widely used by Internet service providers (ISPs) today.

IP addresses are a series of numbers. Some could argue that we are very accustomed to using numbers within the switched network; however, we have to look up the numbers of people we want to call (except for those numbers we call frequently, which we can memorize). Could you imagine having to enter an IP address anytime you wanted to visit a company's Web site? You would first have to look up the company's address, and you would have to know the machine address you wanted to connect with.

Sound familiar? This is the model created for the world's telephone networks. If you want to reach someone, you have to know the number or look it up in the directory. The Internet model was designed to be much more user friendly, and this concept is now being extended to voice services.

The IP address format is dependent on whether or not IPv4 or IPv6 is being used. Since IPv6 is what IMS calls for, we will only discuss IPv6 addressing. IPv6 extends the number of possible IP addresses supported by IPv4 significantly, which is why it is well suited for IMS. Imagine every device connecting to the network requiring its own unique IP address. We would quickly run out of addresses using the current IPv4 model.

However, IPv6 was chosen for other reasons. Security is still a major concern, and IPv6 addresses security through encryption (IPsec). This was another driving factor to standardizing on IPv6 for IMS implementation. IPsec is used to encrypt communications between devices within a secure domain. This prevents eavesdropping on the network from internal sources.

An IP address identifies the interface being used for a connection, and not necessarily the entire node. A node may have multiple IP addresses associated with it, depending on the function of the node. IPv6 addressing takes the form shown in Figure 4.1.

```
ABCD:EF01:2345:6789:ABCD:EF01:2345:6789
```

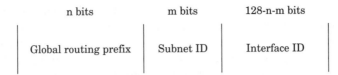

Figure 4.1 IPv6 address format

This example shows that there are eight parts to an IPv6 address, each separated by a colon. Each part contains a 16-bit number, expressed in binary or hexadecimal. Any leading zeros can be eliminated, but trailing zeros change the value for that part of the address if excluded, and therefore, any trailing zeros must be included. An example of a unicast IPv6 address might be:

2001:AC9:800:0:0:EA2:0:4201

IPv4 addresses may be embedded within an IPv6 address where interoperability is needed. In this case, the address notation might look like:

2001:AC9:800:0:0:127.14.32.1

Notice that the IPv4 address is at the end of the string, using the period between address sections rather than colons. The IPv4 address in this example then is 127.14.32.1. Nodes supporting IPv4 recognize the format of the IPv4 address and drop the preceding portions of the address.

There are different types of IP addresses that can be used:

- Unicast address
- Anycast address
- Multicast address

The unicast address is the address assigned to a device interface. This is the commonly used address to reach a subscriber's IP device. An IPv6 unicast address is 132 bits in length, commonly expressed in hexadecimal format when expressed in text strings.

When a device connects into the network, it will receive a Unicast address for the connection. Keep in mind that in reality this identifies the connection in the network to be used to reach the device it is assigned to. All communications intended for the device are sent to the assigned Unicast address.

An Anycast address is used where multiple nodes need to be addressed with the same address. A good example of this would be a P-CSCF or I-CSCF. Rather than assign different IP addresses for each node in the network, by using Anycast addressing, an operator can configure one address for routing to all the P-CSCF and I-CSCF functions in the network. These would then be the addresses used by other operators to route to the IMS. An Anycast address will only route to one node, however, despite the fact that it is assigned to many nodes. The closest node to the originator is where the message is routed to, using rules within the routers to determine distance. When a device needs to connect into an IMS, it can use an Anycast address and be connected to the closest P-CSCF, without having to know the address of that individual P-CSCF, and without having to query another server for the address.

Using Anycast addressing at the CSCF level will certainly eliminate the maintenance of addressing within the network. There may be some security concerns using Anycast addressing, since there is only one address to learn to gain access anyplace in the network, but this can easily be resolved through other security measures as discussed in Chapter 6.

A multicast address is much like a broadcast address (see Figure 4.2). In fact, multicast has replaced the broadcast addresses used in IPv4. With a multicast address, multiple nodes are reached using the same address. Of course, the address is formatted in such a way that all receiving nodes understand that they are receiving a message sent to a multicast address.

Multicast addresses can be used to send communications to all subscribers within an operator's network, for example. Use of this form of address provides an easy method for reaching a broad range of users or devices without the assignment and creation of distribution lists.

There are many other variations for IPv6 outside the scope of this book, but we have covered the main formats used to identify interfaces used for various connections. These are indeed the lowest form of addresses from the IMS sense (of course machine addresses are the absolute lowest identities, but we will skip those). We are discussing IPv6 addresses here, because they do appear in the SIP messaging used along with higher-level SIP identities.

To further complicate addressing, the subscriber could move locations. Mobility is one of the key factors that make VoIP such an attractive offering to many who travel or use communications in a number of different locations. Our society has changed from one that requires stability in a fixed number (with a fixed connection) to one that requires mobility, even if the connection is not fixed.

Since IP addressing is dynamic, it is difficult to use IP addressing to reach subscribers in any form of IP network. Certainly if your IP address were known, anyone could reach you by your IP address. If your IP address is dynamically assigned, there is no means of providing that information to anyone.

Not to mention that IP addressing is not user friendly, and even though we are a society accustomed to memorizing numbers (i.e., telephone numbers) we frequently use,

Figure 4.2 Multicast address format

this no longer makes sense in the IP world. There are much better ways to reach users in today's modern networks.

To accommodate addressing within the Internet, the concept of Universal Resource Identifiers (URIs) were established. This is the form of identity used within the IMS as well. Subscribers no longer have numbers identifying them within the network. These URIs serve as their identities in the same fashion as e-mail addresses today.

SIP Identities in the IMS

Perhaps the most important aspect of the IMS (or any other service network for that matter) is the identity of its users. Our identities are what enable us to communicate freely with other network users, and how we advertise to others how to reach us. But aside from reaching users, identities are important to the network provider as well.

Without the ability to identify users, there is no ability to identify usage and apply billing, by which operators earn their revenues. There is no way to ensure the subscriber accessing your network is authorized to use the services you offer. Identity is important for billing and accounting, security, and even marketing intelligence.

SIP and the IMS can only identify a subscription, since there is no means to identify and authenticate the actual person or persons using the identity (short of some biometrics on the access device authenticating the person). Therefore, identity within the IMS is based on a subscription and not necessarily an individual. Of course, there will be an individual (or individuals) associated with a subscription, and they will be responsible for payment for services rendered, but this does not prevent fraudulent access to a subscription.

Subscription identities are used throughout the IMS for virtually every transaction that takes place within the IMS. Beginning with the registration process, and including each and every request, subscription identity is a crucial and mandatory part of SIP messages within the IMS domain (and usually within VoIP domains as well).

In fact, the user identity provides much flexibility in the offering of services within an IMS implementation. A service provider can provide one subscription to a subscriber but offer many different public identities under that one account. The subscriber is then able to use multiple identities for different applications.

For example, a subscriber may want to maintain one identity strictly for e-mail communications, while using another identity for work-related communications. The subscriber may wish to use yet another identity for personal use, such as gaming and Web browsing. Each identity can be tracked by the service provider with usage captured for each service the subscriber uses with each identity.

The service provider can also offer more flexible rate plans by building service profiles for each of the identities. For example the personal identity may use e-mail and instant messaging, while the work identity only uses e-mail and voice services. The rate plans can be tailored to a subscriber's individual need based on their various identities.

The problem with VoIP implementations has been the lack of being able to validate identities. There are many ways to obtain services for free through the use of fraudulent identities. The SIP message itself is not reliable for identifying the true user identity, which is why the 3GPP added extensions to the SIP protocol for asserting the identity of a device.

The identities (both public and private) are embedded within the device itself, in a fashion that cannot be accessed and modified by the user. This is a new requirement defined by the GSM Association and the 3GPP to thwart fraud in wireless networks.

The identities are part of the SIM card inserted within the device. The SIM application allows users to purchase various devices and take their identities with them no matter what device they are using. When subscribers connect the device to the network, their identities are exchanged with the network as part of the registration process.

Likewise, they are also challenged by the network during the authentication process, and therefore authentication and cipher keys are also stored with these identities. These are also exchanged during the registration process so that the network can verify a subscription when the device accesses the network. These are lessons carried over from GSM implementations over the years.

This is where VoIP implementations failed for many operators, because the protocols were simply not robust enough to ensure that every user was authenticated prior to being able to access services. Security is discussed in more detail in Chapter 6.

Also part of the identity and registration process is the IP address. The subscriber device will provide the assigned IP address along with its identity. This then allows the network to "follow" wherever the user is connected, mapping the URI to the assigned (and dynamic) IP address.

Domain Names and URIs

In the SIP domain, subscription addresses use the form of a Universal Resource Identifier (URI). This is analogous with the Universal Resource Locators (URLs) we use to reach Web sites on the Internet, but they are assigned to subscriptions for reaching individual subscribers.

It is the concept of the URI that makes communications models in the IMS unique. The ability to reach a subscriber based on these very personal identities rather than numbers, and to apply these addresses to all forms of communications, is the purpose of the SIP protocol within the IMS.

A URI can take two forms. A SIP URI uses the same form as an e-mail address, consisting of username@domain. The first part of the address is typically the username of the subscriber, while the last part is the domain name of the network provider where the subscription resides. The last part of the domain name (.com, .org, etc.) defines the type of organization according to Internet rules. The following are examples of the many different domains that are supported:

- .com = commercial
- .edu = educational institution
- .int = international organization

- .gov = government
- .net = network provider
- .org = not-for-profit organization
- .mil = military
- .ca (Canada)
- .de (Denmark)

These are all defined by the Internet Engineering Task Force (IETF). SIP also supports addresses in the form of telephone numbers, referred to as TEL URIs. A TEL URI uses the same form as a SIP URI, substituting the telephone number for the user name. These are used most commonly when a call is originated in a non-SIP domain, or when a call is being placed to a non-SIP network (such as a call from the IMS to a wireline subscriber in the PSTN).

The TEL URI is a good example of interoperability between legacy PSTN and IMS. Since legacy networks will continue to support the use of telephone numbers for some time to come, there remains a need to translate these telephone numbers into public identities for use within the IMS.

The actual conversion process is provided through a function known as ENUM. This function translates an E.164 telephone number into a SIP or TEL URI. The ENUM function does not translate the identity into an IP address, however. This remains the function of the Domain Name Server (DNS). We will talk more about these functions later.

Eventually, TEL URIs may disappear completely as we become more and more accustomed to using SIP URIs for all communications. This is still a long way off, however, and something we may not see for decades.

A subscription can have multiple identities as mentioned earlier. For example, I may have one user identity for my e-mail, while maintaining a telephone number for my cell phone. I may have yet another identity for business purposes, and one for personal use. This is somewhat analogous with screen names with an ISP, where one account can possess several screen names, yet all are billed under one subscription.

In today's networks, you use an ISP to provide e-mail service, and a telephone service provider for wireless and wireline services. This is rapidly changing as service providers are quickly moving to provide all services under one subscription, including video service.

If you have an e-mail address, you already have a URI. The domain name identifies who owns your subscription (your home network). For example, my personal e-mail address is travisruss@aol.com. AOL owns the domain and is my e-mail provider. As the networks evolve, everything you do may be under one domain name.

Still, these must be converted to an IP address so that a connection to the subscriber can be made. There are two methods used for resolving URIs to an IP address. We will talk about network entities first, and then subscription identities.

Each of the IMS entities is assigned a SIP URI, just as subscribers are. The network entities are static, and therefore it is quite easy to automate the function of resolving

URIs to IP addresses. This is accomplished through a rather remarkable database known as Domain Name System (DNS). DNS is a distributed database, consisting of thousands of name servers deployed all over the world, in a truly hierarchical network.

The top level of the hierarchy consists of 13 DNS servers, distributed all over the world. There are 13 of these servers located in the U.S., 1 in Japan, 1 in London, England, and 1 in Stockholm, Sweden. These servers identify the locations of name servers for each of the domains. For example, the servers in the U.S. provide name resolution for all domains located in the U.S. This includes all domains with an extension of .com, .gov, .edu, .mil, etc. The right-most part of the domain name (.com for example) identifies the type of entity that owns the domain name.

Countries are also identified by the right-most portion of the domain name. For example, a domain in Canada would be identified by the .ca as the right-most portion of the domain name. The next portion of the domain name would identify the next level of the name. There can be 127 different levels, but it is rare to see more than four levels used.

This next level is also referred to as registered domains, and they can be purchased through companies acting as registrars for domain names. These registrars manage various domain names for a fee, working with the master keeper of the main list, Network Solutions.

Virtually anyone can own their own domain name for a small fee. The various domain name administrators available through the Internet already offer personalized domain names of individuals and small businesses. The master repository for all domain names must then administer all domains globally.

In the IMS, domain name servers are an important entity just as they are within the Internet. The address of the P-CSCF, I-CSCF, and S-CSCF must be resolved from a SIP URI. The same is true of other IMS network entities such as the Home Subscriber Server (HSS) and Application Servers (AS). Each of these entities is given public and private user identities in the form of a SIP URI, along with an IP address. The domain name server resolves these URIs to the actual IP addresses.

It should be noted that the IP addresses of these entities are static, and therefore changes made to their addresses are infrequent. This is not the case for subscribers and their addresses, so we will talk about that separately. So for the purposes of IMS identities, each identity starts with a domain name, which must be resolved to a physical address in the network. The physical address is an IP address.

Each network uses its own domain name server (or series of servers if it is a large network) to resolve the addresses of its own entities. These servers know the addresses within their domain, but they also have connectivity to the next level of domain servers outside of the network domain, allowing them to query other servers for addresses in other networks. There is no one DNS that knows the addresses of every subscriber in the world.

For example, a network in the U.S. may need to connect to an I-CSCF in Australia. The network operator probably would not have the IP address of the I-CSCF it needs to connect with in Australia, but it does know the address of a domain name server in Australia that would. It would then query this server, which in turn could query yet another server, or simply forward the request to the domain name server with the proper database and IP address being queried.

If a subscriber is outside their home network (they are roaming in another network), the visited network is not responsible for resolving the addresses. The visited network may access its own domain name server to learn the IP address for the I-CSCF of the subscriber's home network, but it would not query a DNS to learn the IP address for the S-CSCF in the subscriber home network. This is left to the I-CSCF in the home network.

For subscribers, the use of a domain name server is not as important, because the domain of the service provider is all that is needed. To locate the address of any individual subscriber, networks would first query the registrar in the home network, which is the S-CSCF. The S-CSCF stores the IP address of each subscriber during the registration process.

As a subscriber moves about the network (and changes his or her IP address) the S-CSCF is updated with the user's identities and associated IP addresses. Note that a device may have multiple IP addresses assigned to multiple user identities as well.

User Identities

User identities are used for two purposes in the IMS. First and foremost, routing of requests and responses are directed to a user identity, which is associated with an IP address (disclosed during the registration process). This same identity cannot be used for billing, however, as this would pose a security threat.

For this reason a separate identity is used for billing and accounting within the IMS. This identity is not disclosed outside of the network and is not visible to end users of IMS services. It is only visible to network operators that assign them. There are two different types of user identities; a public user identity and a private user identity.

Private User Identity   The private user identity is assigned by the home service provider. This identity is used for registration of the subscriber, authentication, and authorization, administration of the subscription, and of course billing and accounting. The private user identity is never used for routing. It is embedded within the device itself, either through some administrative effort or some automated function, dependent on implementation.

For example, earlier in this chapter we discussed the use of SIM cards within a device. Regardless of whether the device is a wireline or wireless device, the SIM card would be inserted into the device. The SIM card is programmed by the service provider with the private user identity for the subscription. Think of the private user identity as the identity for a subscription.

If there is no private user identity assigned in the case of wireless, then the International Mobile Subscriber Identifier (IMSI) is used to create the private user identity. The private user identity would take the form of *<imsi>*@mnc*<MNC>*.mcc*<MCC>*.3gppnetwork.org. The first part of this address is the IMSI itself, followed by the mobile national code (MNC), and the mobile city code (MCC). This is then a TEL URI format.

This private identity is not advertised outside of the home network and is not visible to the end user. It is used within the home network only and usually consists of the subscription identity and possibly an account number. It identifies the subscription itself and is associated with at least one, but most likely several public user identities.

Whenever a subscriber registers in the network (by activating his or her phone for example), the device must send the private user identity, which is embedded within the device (usually in a SIM card). The device also sends the IP address it was assigned when it connected to the network. Only one private user identity can be stored in the device (however, a device and private user identity may have multiple public user identities associated with it).

The private user identity is assigned one time when a subscriber starts an account with the service provider. It is stored in the service providers Home Subscriber Server (HSS) along with a profile of services subscribed to, and current registration information. The only time a private user identity is authenticated is during registration.

Public User Identity   The public user identity is used for routing. This is the identity subscribers use to advertise how they can be reached. A telephone number in the form of a URI is considered a public user identity (for example, +9193681397@tcg.com is a public user identity in the form of a TEL URI).

The public user identity can also be an e-mail address (such as travis.russell@tekelec.com), or any other form of SIP URI, and is stored on the device. The device stores identities in an application that resides on the Universal Integrated Circuit Card (UICC). Look inside any GSM phone and you will find a tiny circuit card that inserts into the phone (commonly referred to as the Subscriber Identity Module, or SIM).

If there is no public user identity provided, the network then uses the IMSI in the same form it did for the private user identity. The IMSI form is not used for human consumption. It is only used by the network to determine how to route the call to the home network and reach the HSS. The public (and private) user identities are then learned from the HSS. The IMSI form would take the form of:

```
sip:imsi@mnc<MNC>.mcc<MCC>.3gppnetwork.org
```

The IP address for the device is assigned dynamically, by the network provider the device is connecting with. When the device is activated, the device is given an IP address by the network, which it then stores in memory. This IP address is then sent along with the public and private user identities to the serving S-CSCF for storing during the registration process. The S-CSCF sends this information to the HSS, where it is also stored for the duration of the registration.

Again it is this concept of using user identities in the form of URIs that helps support mobility in the IMS. All that is needed is a subscribers identity, and calls (or e-mails, or text messaging, or any other form of communications) can be sent to the subscriber no matter what IP address that subscriber has been assigned. Only the network needs to know the IP address for the subscriber at any point in time.

It should be noted here that there is no DNS query for subscriber identities. The URI for a subscriber is resolved to an IP address by the S-CSCF and the HSS functions.

The IP address is updated anytime the device changes connections and re-registers with the network. This would be the case for a mobile subscriber who is roaming in other networks (or changes cell sites within his or her own home network).

The registration process is where the IP address is "bound" to the public and private user identities. The concept was stolen both from the mobile industry, where the concept of registration is used with cell phones accessing the network, and the Internet, where dynamic IP addressing is commonly used.

Only one public user identity can be stored on the device. Other public user identities are stored in the service provider's HSS and are shared with the device during registration. This allows the device to then store the other associated public user identities in memory (but not on the UICC). The SIP (header) *P-ASSOCIATED-IDENTITY* provides this information to the device.

Figure 4.3 illustrates the relationship between the private user identity and the public user identity. Notice that there can be multiple service profiles associated with one private user identity. Likewise there can be multiple public user identities associated with one private user identity.

A good example of how a subscriber can have multiple identities is when a subscriber has multiple devices. Typically, subscribers have a cell phone, an e-mail account, a home telephone number, a Blackberry, a personal e-mail account, and possibly another set of identities used for hobbies or other interests. Each one of these uses a public user identity.

I may wish to have all of my business calls directed to my desk phone at work (using a public user identity consisting of my work phone number). At the same time, I may want to have family and friends directed to my personal cell phone. This is also identified by a unique public user identity.

In the IMS, a subscriber will have the ability to direct calls to different devices depending on who is calling, and when they are calling. This is accomplished through the use of a Presence server, responsible for directing certain calls to certain devices according to preset conditions (such as who is calling and time of day, for example).

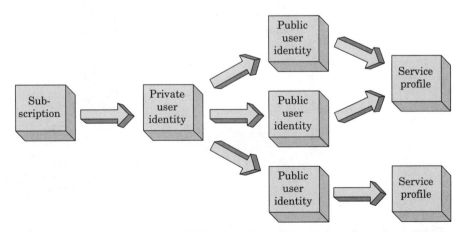

Figure 4.3 Relationship between the private user identity, public user identity, and a subscription

As networks mature and the IMS implementation becomes more wide scale, the network will be able to monitor users' locations based on their movements around the network. For example, the network could determine by tracking the handset movements that a subscriber has left their office and is now commuting. Given this condition, the user identity for work may dictate that all work calls be routed from the user's desk to the user's cell phone.

However, as a subscriber arrives home, the network could sense this change (again based on Presence configurations) and begin directing all work calls to the subscriber's voice mail service. At the same time, the network could direct all personal calls from the cell phone to the subscriber's home phone.

This is a rather unique capability provided by the IMS model and the various call session control functions throughout the network. Working with various application servers, the possibilities are endless.

While a subscriber can have multiple user identities, all identities (both public and private) must be registered within the same registrar (the same S-CSCF). The service profiles associated with each of the identities is also stored within the same S-CSCF (downloaded from the HSS at registration time).

This is paramount to providing ubiquitous service no matter where a subscriber is located. Again this model was taken from the wireless community, where all handsets register their location with their home network so that calls can be routed to the handset no matter where the subscriber is located.

A service profile is also identified through the use of a service identifier. A public service identity is associated with the service profile in the HSS. The public service identity can be shared across multiple devices as mentioned previously, and as illustrated in Figure 4.4.

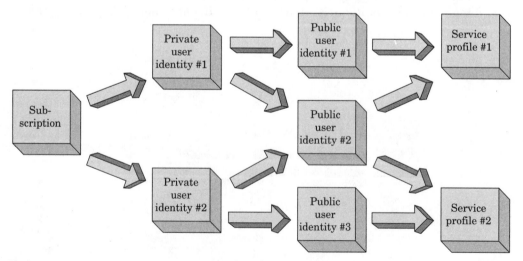

Figure 4.4 Relationship between subscriptions, private and public user identities, and service profiles

As is the case with user identities, services also have public identities. The public service identities are used to access services such as conference bridges and whiteboard applications, as an example. They are then used to identify the services a subscriber is allowed to use within the network.

Service identities are not associated with any user identities other than identifying what services they are allowed to use, as these are actually network services and not part of subscriptions. They are mentioned here as an explanation as to how services are reached through the IMS. Otherwise, it would be difficult to identify specific services that a subscription pays for as part of the subscriber profile.

A service identity can be static or dynamic and is assigned by the application server. The service identity is used within the IMS for routing to the proper application server. When an application server registers in the network, it will advertise its capabilities to the HSS and S-CSCF, along with its service identity. The service identity is also sent in the charging parameters sent through SIP in the IMS domain.

The service identity then becomes a part of the subscriber profile within the HSS. Notice that a service profile could be in multiple public user identities. It simply identifies what service each identity is to be able to access, so even partner networks will be able to access and provide the same service even when the subscriber is outside their home network.

This is yet another important reason for the use of IMS identities and service identities. As subscribers roam from network to network, the visited network needs the ability to access the same application servers and deliver the same services the subscriber uses at home. This is not always possible in today's networks.

In the IMS model, however, partner networks would be able to identify what services the subscriber is allowed to use (through the subscriber profile stored on the HSS) and deliver that same service no matter where the application server resides.

We have discussed a lot about subscriber and service identifiers in the IMS. As we discuss other processes within the IMS, it will become even clearer the value these identities bring to service providers.

5

Establishing a Session in the IMS

In legacy networks, subscribers make voice calls. They use a different network to send e-mails, and yet another network to access the Internet through a wireless connection. They watch videos and television broadcasts through another network such as cable and satellite.

The Internet has forever changed this paradigm. No longer will we need all of these disparate networks in our lives to communicate and entertain ourselves. As operators begin transforming their networks into an all-IP infrastructure capable of supporting all of these media types, all we need is a single connection point.

The IMS brings this possibility to reality. It allows a subscriber to establish a single connection for everything that subscribers need, including voice and video. Once that connection to the network is made, everything he or she needs becomes immediately available through that connection. However, the network still needs multiple sessions.

Everything that a subscriber does to communicate becomes a session in the IMS environment. This is probably the largest benefit of IMS networks. Rather than have disparate networks managing various forms of media, operators can consolidate their networks into one network for all media types, all under one common control.

This means that even e-mail and video sessions are controlled and managed by the same control network, and visibility to the various transactions within the network is much simpler. This is certainly one of the largest holes in today's telecommunications networks. Operators are able to support multimedia sessions, but they lack the controls and mechanisms to effectively and profitably manage those sessions as a whole.

For example, a wireless provider is quite accustomed to delivering quality voice and data to wireless devices even when the subscriber roams into other networks. But the same operator would have difficulty delivering video to that subscriber in their home network, much less when they are roaming. This requires more than control of the session and connections; it requires interoperability of all of the network components delivering the service to the end device.

Interoperability is more than the ability to connect multiple devices to one another. It's about being able to integrate the various services so that they are truly seamless to the subscriber. This means that subscribers need only to activate their device, select the service they want, and begin enjoying the content being delivered to them.

Interoperability also means providing a common infrastructure to support the service, from ordering to provisioning and ultimately billing, again fully integrated with all the other service platforms being used to deliver services to consumers.

This is the spirit of the IMS. It's about more than enabling all these different media types and providing the delivery infrastructure. It's about enabling all of the various entities within the network to work together, seamlessly and ubiquitously.

In this chapter we will look into how subscribers gain access to the IMS network and the registration process. We will then look at session establishment and management. This will give you a better look at how IMS and SIP can be used to control everything that a subscriber does in your network.

Accessing the Network

First of all, access means many different things to a subscriber. It might mean using a PDA with an 802.11b application to access the Web and download music files, or it might mean using a cell phone to make voice calls. It might even mean picking up the phone at home and placing calls, or using a broadband connection to surf the Internet.

The method of access into the network becomes transparent to the IMS, because there is no transport of actual bearer traffic within the IMS. The voice, the data, the video, and everything else being delivered to the subscriber takes a different path, using the access method for delivery.

The IMS is used to simply control all of those connections, and to manage those connections in such a way that services can be provisioned and charged for accurately without multiple systems in place to support those functions. Otherwise, there would be a need for many different protocols and platforms to deliver true multimedia to subscribers.

In the IMS, any time a subscriber wishes to establish any form of a dialog, they must create a session. Think of a session as a portion of a dialog between two parties. For example, in a videoconference, the voice portion of the transmission would be one session, while the video portion would be a completely different session.

This allows networks to use different media types to deliver the bearer traffic. The media types and the path of the bearer traffic then become transparent to the user. The network only needs one simple method of establishing and managing all of these connections within the network using a common call control layer. The purpose of the IMS, as we have already said many times, is to simply provide this control layer.

The IMS uses SIP for controlling all sessions within the IMS domain; therefore, the rules of SIP registration and session establishment apply here. There are some extensions to SIP that have been defined by 3GPP specifically for use within the IMS domain, in an effort to make IMS more robust and secure.

This is because the early versions of SIP were developed for use in delivering voice over the Internet. The concept of secure communications over the Internet could be a bit of an oxymoron in the "Wild West" of this space. The Internet has always used minimal restrictions on the users in terms of connectivity and security.

The 3GPP has recognized that SIP, according to RFC 3261, did not provide the robust controls needed for their wireless networks, and therefore added many enhancements and extensions to the protocol for use in traditional telecom models.

The first step for any session establishment is establishing a connection to the IMS network. This does not mean connecting to the IP network, or the GSM network, but the call control network. The subscriber must first establish the physical connection into the access network (such as a landline or a wireless connection). Then they can begin communicating via SIP to the IMS.

Depending on the device type, and the location of the subscriber, there may be direct access into the IMS or the connection may rely on various gateways and other networks to gain access. For example, if a subscriber has a SIP-enabled device, the service provider may be able to provide a direct connection to the P-CSCF within their network (the P-CSCF function is described in greater detail later). If not, then the subscriber's device must connect through a media gateway controller, which can convert the signaling (call control) to SIP for communicating into the IMS. Following are several examples of how this is accomplished.

In a wireline network, if a subscriber is using a PC or other stationary device to access the IMS, the device must first secure an IP address. This process is started once the device is activated (turned on). If the PC is connected to a local area network (LAN), the LAN would assign the IP address (or the address could be hardwired to the location). In a service provider's network, the connection may be through a DSL or ISDN connection, where the network then provides the IP connectivity and the IP address is assigned dynamically. The mechanics for how the device is assigned an IP address is really outside the scope of this book, so we won't go into details about dynamic address assignment.

Once the device has obtained an IP address, it must then learn of the closest IMS network access. This is assuming that the device supports SIP and IMS procedures. If the device does not support SIP, then interworking procedures are used to connect.

Interworking means using processes such as media gateway controllers (MGCs) for converting from SS7 (for example) to SIP. The MGC is then responsible for managing the connections on both sides of the MGC. The interworking between other networks is critical to ensuring that legacy networks can be left in place while IMS is implemented. Again, only the call control is sent through the MGC function. In fact, the MGC communicates with media gateways (MGs) to control the actual voice transmission. The MG supports the bearer traffic under the control of the MGC.

Locating the P-CSCF

As already mentioned, a device connects first to the access network. Once this connection is made, the device needs to communicate its location (address) to the network so that it can begin receiving e-mail, voice mail, phone calls, etc. This requires connecting to the IMS.

Access to the IMS network is achieved through the Proxy-Call Session Control Function (P-CSCF). The device must find the P-CSCF within its present domain. P-CSCF discovery can be coupled with the IP address assignment (whereas the IP address and the P-CSCF location are delivered together), or through a query to a Dynamic Host Configuration Protocol (DHCP) server. The DHCP should return the domain name for the appropriate P-CSCF serving the area the device is located in (or serving the domain the device is located in).

For example, if I power up my PC at home and connect to the network, the network will provide me the address of the P-CSCF that is serving my area. The address consists of a domain name only. To obtain the actual IP address, my PC will have to query a Domain Name Server (DNS). Once my PC learns the IP address for the P-CSCF serving its area, the device can begin registration procedures with the IMS network. No sessions can be established until registration has been completed.

This is much different than the way VoIP has been implemented today. There is a registration process in some implementations, but communications with VoIP networks do not require much in the way of authentication and authorization. In other words, registration is much simpler in most VoIP networks, only requiring the communication of the device's location so that call routing can be completed.

For a wireless device, the process is somewhat different because the device can be outside of its own home network (roaming). When this is the case, the device must still obtain an IP address, as well as learn the address of the P-CSCF serving the area in which the device is located. The difference is that the device may be locating a P-CSCF from another service provider rather than its own home network.

As I have discussed in Chapter 1 and 2, the P-CSCF serves as the access point to the IMS and therefore controls all communications from other networks. The P-CSCF serves as the gateway into the IMS, communicating to all other SIP devices. All subscribers must connect first to the P-CSCF prior to accessing the S-CSCF.

The P-CSCF provides some protection to the S-CSCF and the HSS. This means that the P-CSCF will do some screening to ensure that the subscriber trying to communicate with the network is registered and authorized to access the network. It does this through security associations.

Once a device has been authenticated and registration has been completed, the P-CSCF assigns secure ports for the device to use for all subsequent communications. Port 5060 and port 5061, typically used for SIP communications, are only used for registration. The P-CSCF discards any *INVITES* or other SIP message types received on these ports.

This ensures that communications into the network are from trusted devices, and not from unauthorized devices. This is one of several different methods used for security within the IMS, but it's not by any means the only measure. This is one of the primary methods used by the P-CSCF to protect the resources of the HSS and the S-CSCF.

If you are located in a non-SIP network, other devices are used to provide the gateway function between the IMS and legacy networks. We have already talked about some of these devices in earlier chapters. The media gateway control function (MGCF) provides the gateway function between the SS7 network and the SIP IMS network, but it is providing more than simply protocol conversion. The facilities connected to the media gateways (MGs) under the control of the MGCF are managed as well on the TDM side of the network using the SS7 protocol, and therefore the MGCF must be a stateful entity.

Now that we have a basic understanding of how access to the network is accomplished, we can start digging into the processes used to register.

Registration in the IMS Using SIP

This is probably a good time to talk about the difference between subscribers and subscriptions. Keep in mind that throughout this book when we talk about subscribers and/or subscriptions, the network cannot identify individuals. The network can only register and authenticate subscriptions, which do not necessarily identify real people (in other words, unauthorized individuals could register using someone else's subscription). We will use both terms interchangeably throughout the book.

In order for the network to connect calls to a device, it must first know where in the network to find the device, and how to route session control messages (SIP) to the device. This is accomplished by forcing all devices to register with the network when they are activated. This registration provides the network the information needed to be able to locate the device, and route SIP messages to the device, as long as it remains activated.

This also means that as the device changes locations and associated addresses, the device must re-register with the network to provide its updated location information. The re-registration process does not have to be as stringent as the initial registration unless the service provider chooses for it to be.

When a subscriber is registered, the subscriber registers his or her private and public user identity with the network. This is so the network knows how to route calls to the subscriber, associating the private and public user identities with an actual IP address. However, it is possible for a subscriber to have many public user identities. For example, you may have an identity for home (your home phone number for example), while maintaining a separate identity for work. Each of these public identities is registered with the network either simultaneously or through separate registrations.

If you have several different devices, such as PDAs, cell phones, Blackberries, and home PCs, you may use a different identity for each one (or the same identity). As each of these devices connects to the network and receives a physical address, it will then register this address along with the chosen identity. This is how subscribers could end up with multiple devices and multiple identities.

Only one public user identity is stored within the S-CSCF at a time. All other public user identities are stored in the service profile on the HSS. It is also possible to have multiple public user identities registered with multiple devices (and therefore multiple IP addresses). This information is also stored within the HSS, while each unique device address (IP address) is stored with its associated public user identity in the S-CSCF.

All devices attempting to register will be challenged by the network registrar (the S-CSCF assigned to the subscriber during the registration process). This challenge is to authenticate the subscription, preventing unauthorized access. The registration is challenged by rejecting the initial registration and forcing a second registration message containing the proper credentials.

The registration is rejected by returning the response 401 Unauthorized to the device. When the S-CSCF sends this response, it stores the *CALL ID* from the *REGISTER* and queries the HSS for the subscriber's credentials. The credentials consist of the cipher key as well as the authentication key. These are explained in much more detail in Chapter 6.

When the subscriber device receives the response, it will generate a new *REGISTER* containing the proper credentials and with the same *CALL ID* as the original *REGISTER*. The S-CSCF then knows by the *CALL ID* that this is in response to the earlier rejection and can then correlate the two *REGISTER* messages.

Another method would be to always have the device send its credentials each time it registers a new location with the network. This would work but would also generate a lot of traffic between the S-CSCF and the HSS. There are cases when a device is sending a re-register that the S-CSCF would not have to query the HSS again because the device is already known to be trusted.

Only one registration can be executed at a time. The device cannot send a *REGISTER* while waiting for a response from a previous *REGISTER*. One registration must be completed before starting another one.

Registration must be changed whenever the device changes locations. For example, when a mobile subscriber is roaming, that subscriber will have to register each time he or she roams into a new service area (or changes cell tower connections). Remember that the registration process is for informing the HSS and the S-CSCF of the present location (physical address) for the device so that requests and responses can be correctly routed to the device.

Registration is also a means for authenticating subscribers and verifying their credentials. This is accomplished by equipping devices with an authentication key that can then be stored in the Home Subscriber Server (HSS). Each time the subscriber registers, the device will be challenged by the registrar (the S-CSCF) to provide the authentication keys. Chapter 5 discusses more about the authentication process and overall security in the IMS.

One final note about authentication. The *AUTHORIZATION* header in the *REGISTER* message may contain the parameter *INTEGRITY-PROTECTED=YES*. When this occurs, authentication is not required, because the subscriber has already registered and has been authenticated already. If this parameter is set to NO, then the registrar should challenge the subscriber device and authenticate the subscription.

This is implementation specific, as there are some service providers who may not wish to use this capability. Indeed some service providers may wish to authenticate each and every time, to ensure there are no registration hijacks, for example. The IMS standard supports either method.

Once the device has connected to the network and has located the P-CSCF, it can send a SIP *REGISTER* message. The *REGISTER* message will contain the domain name of the subscriber (this identifies the home network that the subscription belongs to).

The domain name is stored in the device, usually in the IP Multimedia Services Identity Module (ISIM) as part of the user identity.

The ISIM is an application that runs on the device's Universal Integrated Circuit Card (UICC), or what is more commonly known simply as the SIM card. The ISIM identifies the subscriber by providing a private user identity and one or more public user identities. These are used for the registration process and for authenticating the subscriber.

The ISIM also contains the authentication key and the cipher key. These are also exchanged during the registration process after the first registration is challenged. The authentication and cipher keys allow the device to calculate proper credentials that should match with those stored on the HSS and the S-CSCF.

Following are the headers and parameters used in the *REGISTER* message:

- *REQUEST-URI* populated with the subscriber's public user identity and domain name of the home network
- *FROM* header populated with the public user identity being registered
- *TO* header containing the public user identity being registered
- *AUTHORIZATION* header populated with the private user identity being registered
- *CONTACT* header populated with the subscriber's IP address for direct routing of messages
- *VIA* header populated with the IP address or domain name of the device initiating the registration
- *EXPIRES* header set to 600,000 seconds
- *SECURITY-CLIENT* header identifying the security methods supported, the IPsec algorithm, and security association parameters

The domain name will be used to identify the Interrogating—Call Session Control Function (I-CSCF) for the subscriber's home network. The P-CSCF must resolve the domain name through a DNS query to obtain the actual IP address of the I-CSCF.

The P-CSCF will add several headers and parameters to the *REGISTER* prior to forwarding it to the I-CSCF:

- *PATH* header populated with the P-CSCF address
- *REQUIRE* header populated with the option tag *PATH*
- *P-CHARGING-VECTOR* header populated with the *IMS CHARGING IDENTITY (ICID)*
- *AUTHORIZATION* header populated with the *INTEGRITY PROTECTED=YES* if the registration was received through secure procedures and *NO* if not
- Remove the *SECURITY-CLIENT* header and store it if the registration was received without integrity protection
- *P-VISITED-NETWORK-ID* header populated with the visited network domain name

The I-CSCF serves as the access point into the subscriber's home network. When the I-CSCF receives the registration request from the subscriber, it will query the Home Subscriber Server (HSS) within the home network to determine if a Serving–Call Session Control Function (S-CSCF) has already been assigned for this subscription. The S-CSCF serves as the registrar for the IMS.

If a S-CSCF has not been assigned yet, the I-CSCF will assign the S-CSCF according to local policy. The S-CSCF can be assigned in terms of geography or of session requirements, whichever the operator chooses to implement. More on that later.

To locate the proper HSS, the I-CSCF may optionally use the Subscription Locator Function (SLF). The SLF provides the HSS addresses according to the subscription URI for each network. The SLF maps the subscriber identity to the assigned HSS and forwards the address of the assigned HSS to the I-CSCF. If the SLF cannot resolve the URI to an HSS, the I-CSCF will fail the request and return the response 403 Forbidden to the subscriber. Use of the SLF is operator dependent.

Basic Session Registration

The I-CSCF is also responsible for assigning the S-CSCF to the subscriber. Assignment is based on one of two things: either the location of the subscriber or the capabilities required for the session. One approach is to assign the S-CSCF according to capability and availability. If the subscriber has not yet registered, the I-CSCF will look at the required capabilities (for example, will video have to be supported?) and S-CSCF availability. The S-CSCF that supports the session requirements and is available is then assigned to the subscriber. When the subscriber changes locations and registers again, that subscriber may very well be assigned to a different S-CSCF. Other criteria can be used by operators for assignment of the S-CSCF.

Some operators may choose to assign the S-CSCF by geography (much as telephone networks are designed today). If this is the case, then all S-CSCFs within a network domain will support the same capabilities and be assigned according to a subscriber's location during registration.

This is the more likely scenario for traditional operators, since their network resources are already deployed according to geography. This also makes it simpler to maintain the network in the long run.

Once the S-CSCF identity has been established, the *REGISTER* message is forwarded to the S-CSCF for registration. The S-CSCF stores the address of the P-CSCF so that it knows how to send any responses back to the subscriber. The S-CSCF will first challenge the subscriber by returning a response 401 Unauthorized. The S-CSCF will then query the HSS for authentication data. The S-CSCF only challenges a subscriber twice. Likewise, a user device will only respond to the 401 Unauthorized response twice.

The HSS stores both a cipher key and an authentication key. Both of these keys are used with specific algorithms to calculate the subscriber credentials and to identify how to decipher encrypted messages. These keys are passed to each of the entities involved in the registration (the S-CSCF and the subscriber device specifically).

When the device receives the response 401 Unauthorized from the S-CSCF, it will then send another *REGISTER* message containing its credentials based on the authentication and cipher keys stored in the ISIM application. These keys cannot be changed by the user and will be used each and every time the subscriber accesses the network. They are only known to the service provider that assigns them and the device itself. The new *REGISTER* message will use the same path as the first message, per IMS SIP routing rules.

The *REGISTER* message will contain the P-CSCF address and network identifier (usually in the form of a domain name), the public user identity to be registered, the private user identity assigned to the subscriber, and the device's IP address. The *CALL-ID* value will be the same as the initial *REGISTER* message so that the S-CSCF knows that this is a response to a previous challenge.

When the S-CSCF receives the *REGISTER* message, it identifies the subscriber by examining the *TO* header populated with the public user identity and the *AUTHORIZATION* header populated with the private user identity. If the S-CSCF verifies the authentication keys are correct, it will send a response 200 OK to the subscriber. The S-CSCF routes the response using the same path as the request. The response 200 OK will be populated as follows:

- *PATH* headers indicating the addresses used to reach the S-CSCF

- *P-ASSOCIATED-URI* containing all the public user identities the user is authorized to use for this subscription

- The default public user identity listed as the first entry in the *P-ASSOCIATED-URI* (as identified by the HSS)

- *SERVICE-ROUTE* header populated with the address of the S-CSCF assigned to the subscriber as well as the I-CSCF address

- The address of the I-CSCF providing topology hiding populated as the first address in the *SERVICE-ROUTE* header if applicable

- The *P-CHARGING-FUNCTION-ADDRESS* header provided by the HSS if the P-CSCF identified in the *P-VISITED-NETWORK-ID* header is in the same network as the S-CSCF

If the S-CSCF does not respond to the *REGISTER,* or if it returns the response 3xx or 480 Temporary Unavailable, the I-CSCF will select another S-CSCF (depending on operator policy). If there are no available S-CSCFs to receive the request, then the I-CSCF returns the response 600 Busy Everywhere.

When the P-CSCF receives the response from the S-CSCF (through the I-CSCF) the P-CSCF stores the *SERVICE-ROUTE* header in the order received and uses the addresses in the headers to establish a routing list to be used for all subsequent requests. This list is associated with the public user identity within the P-CSCF. The P-CSCF also stores all of the public user identities identified in the *P-ASSOCIATED-URI* header in the response.

This is an important security function of the P-CSCF. Should a registration become hijacked and a perpetrator sends an *INVITE* or other message using the identity already registered, the message will be routed using this routing information regardless of the SIP headers. The P-CSCF will change the routing of the request to follow the route as registered. This prevents someone from hijacking their service from another point in the network.

The *SERVICE-ROUTE* header is used in the response to a *REGISTER* (200 OK). The values are derived from the *PATH* header contained in the *REGISTER* message and identify the addresses of all the network elements used in the path of the message. This is the only time this header is used.

Upon receipt of the response 200 OK from the S-CSCF, the device will extract the following parameters from the response and store them for session establishment:

- Expiration time for each public identity
- List of URIs found in the *P-ASSOCIATED-URI* header
- The first URI in the *P-ASSOCIATED-URI* as its default public user identity
- *RECORD-ROUTE* headers for a route list
- Security association lifetime to either the previous value or 30 seconds longer than the *EXPIRES* value

The information sent over the Cx interface shown in Figure 5.1 includes the public user identity, the private user identity, and the P-CSCF network identifier. This information is then stored in the HSS as part of the subscribers profile.

When the HSS receives this information, it must first determine if the subscriber is already registered through another S-CSCF. If the subscriber is not registered, the HSS checks the service profile to verify the subscriber is allowed to roam in the network identified by the P-CSCF domain name.

The HSS then returns a response back to the I-CSCF containing the name of the S-CSCF if previously registered. If not, then the HSS returns the capabilities for the subscriber. This is an indication to the I-CSCF that it must assign the S-CSCF. The I-CSCF then forwards the *REGISTER* to the newly assigned S-CSCF.

Notice in Figure 5.1 that the S-CSCF and the HSS communicate once the *REGISTER* has been received by the S-CSCF. This is so the S-CSCF can obtain any additional addresses registered for the subscriber, as well as any security information. The HSS stores the S-CSCF address for future queries from other network entities (such as during session establishment). The S-CSCF may also identify any networks the subscriber may be roaming in (identified in the *P-VISITED-NETWORK* header).

Once a subscriber has registered with the network, they must update their registration prior to its expiration. This is based on timers and the *EXPIRES* parameter found in the *REGISTER* message. The device will re-register 600 seconds prior to the expiration time if the initial expiration time was greater than 1200 seconds. If not, the device will re-register when the expiration timer reaches half its initial value. Should a subscriber fail to update their registration prior to expiration, the S-CSCF will deregister the subscriber from the network.

Re-registration follows the same procedures we have already described, for the exception of routing. Since the routing is already established for all messages from the

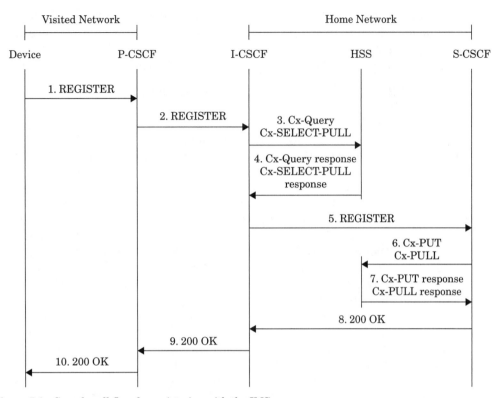

Figure 5.1 Sample call flow for registering with the IMS

device to the S-CSCF (and recorded via SIP headers by each network entity during the registration process), the I-CSCF does not pay any attention to the P-CSCF address. The S-CSCF can also forego the queries to the HSS during re-registration, although this is left to the operators to decide.

When re-registering with the network, the subscriber's device sends the *REGISTER* message with the following headers and parameters:

- *AUTHORIZATION* populated with the private user identity
- *FROM* populated with the public user identity to be registered
- *TO* populated with the public user identity to be registered
- *CONTACT* populated with the SIP, URI, and IP address of the subscriber
- *VIA* populated with the IP address of the subscriber device
- *EXPIRES* set to 600,000 seconds
- *REQUEST-URI* populated with the domain name of the home network
- *SECURITY-CLIENT* populated with the supported security mechanisms

- *SECURITY-VERIFY* populated with the content from the *SECURITY-SERVER* header received in the response 401 Unauthorized during previous authentication and registration

- *SUPPORTED* populated with the option tag *PATH*

- *P-ACCESS-NETWORK-INFO* identifying the type of network used to access the IMS

After registration, if the subscriber should be assigned a new IP address (after roaming into another network, for example), the subscriber must cancel all previous registrations and register with the new IP address. Otherwise, the network will be attempting to connect to the old IP address and sessions will fail.

It is purely up to the service provider whether or not to force the subscriber to authenticate himself or herself again during the re-registration process. Some operators may choose to reject the re-registration as a security measure, forcing the subscriber device to authenticate itself again using the same procedures we discussed earlier.

However, this will result in additional traffic to the HSS as well as through the network. For this reason some operators may select the latter procedure and allow subscribers to re-register without authenticating again.

When a subscriber wishes to de-register from the network, the subscriber device sends the *REGISTER* message again, with an *EXPIRES* header value of 0 seconds. This indicates that the existing registrations for the subscriber should be canceled.

Subscribers de-register when they are disconnecting (powering down) from the network. All sessions in progress are terminated, but any IP connections must be maintained until the subscriber has successfully de-registered with the S-CSCF.

The S-CSCF will then send a *PUT* over the Cx interface to the HSS containing the public user identity or private user identity, along with instructions to either clear the S-CSCF or keep the S-CSCF. If told to keep the S-CSCF, the HSS will maintain an association between the subscriber and the S-CSCF (for use for other services that may not require registration).

There are a number of reasons why a subscriber may wish to de-register, but the operator may also force de-registration (dependent on operator implementation). Reasons for this include

- Network maintenance

- Network or traffic (for example, if two network operators change the parameters of their roaming agreements, they would want to de-register all subscribers to prevent multiple registrations)

- Application layer issues (various applications may require de-registration of individual subscribers or groups of subscribers)

- Subscription management (delinquent accounts, for example)

Throughout the registration processes, the various devices will store specific information during and after registration. Table 5.1 identifies the information stored by each of the devices throughout the registration process (during and after).

TABLE 5.1   IMS Entities and the Information They Store During and After Registration

Device	Before	During	After
	Credentials Home domain Proxy name & address	Credentials Home domain Proxy name & address	Credentials Home domain Proxy name & address
P-CSCF	Nothing stored	Device address Public user identity Private user identity	Device address Public user identity Private user identity
I-CSCF	HSS or SLF address	S-CSCF address and name P-CSCF domain Home network domain	Nothing
HSS	User service profile	P-CSCF domain	S-CSCF address & name
S-CSCF	Nothing	HSS address User profile P-CSCF address P-CSCF domain Public identity Private identity Device IP address	HSS address User profile P-CSCF address P-CSCF domain Public identity Private identity Device IP address

Various network entities may wish to be alerted whenever a subscriber is registered or when he/she has a changed registration, for example, an application server that supports a location-based service (LBS), or a Presence server providing information about the availability and status of a subscriber. These entities would need to update their status about the subscriber and therefore should be notified whenever the subscriber status changes.

This is supported through event notification. The S-CSCF is responsible for sending event notification to any of the application servers (or other entities within the network) that have subscribed to the service.

The subscriber's device also subscribes to Event Notification, in the event the subscriber registers the same public user identity with another device (so the subscriber can receive calls on multiple devices' for example). Subscription to Event notification is achieved using the *SUBSCRIBE* method with the following headers and parameters:

- *REQUEST-URI* set to the value of the public user identity that the device wants event notification for
- *FROM* header set to the same public user identity as the *REQUEST-URI*
- *TO* header set to same public user identity as the *REQUEST-URI*
- *EVENT* header set to *REG*
- *EXPIRES* header set to 600,000 seconds
- *P-ACCESS-NETWORK-INFO* identifying the network type used to access the IMS

The P-CSCF also subscribes to Event Notification and sends a *SUBSCRIBE* to the S-CSCF with the following headers and parameters:

- *REQUEST-URI* populated with the subscriber's default public user identity
- *FROM* populated with the P-CSCF address
- *TO* populated with the default public user identity of the subscriber
- *EVENT* set to *REG*
- *EXPIRES* set to a value higher than what was received in the 200 OK response
- *P-ASSERTED-IDENTITY* populated with the address of the P-CSCF identified in the *PATH* header of the initial *REGISTER* message
- *P-CHARGING-VECTOR* populated with the *IMS CHARGING IDENTIFICATION (ICID)*

The P-CSCF forwards the *SUBSCRIBE* to the I-CSCF in the home network of the subscriber that the P-CSCF is subscribing to. The I-CSCF will then forward the request to the S-CSCF assigned to the subscriber. This allows the P-CSCF to maintain the state of the subscriber and that subscriber's registration.

Anytime there is a change in the registration status of a subscriber, the *NOTIFY* method is used to alert all subscribed entities of the change. For example, the network may de-register a subscriber from the network. This will result in a *NOTIFY* being sent to all subscribed entities, providing a list of all public user identities for the subscription being de-registered.

Thus far we have discussed accessing the IMS network and registration procedures. Accessing the IMS allows devices to quickly establish their location and notify the network of their location. This is assuming that the subscriber device is already SIP capable and compatible with IMS. However, this is not yet the case, and therefore subscribers will have to access the IMS through legacy networks. This means there are additional entities required to convert the call control data to SIP for IMS compatibility.

This next section describes the procedures through other non-IMS networks for gaining access and establishing sessions. We will then explore how a session is established and managed using SIP.

Interworking with the PSTN

As mentioned, operators will depend on their existing legacy networks to provide services to their subscribers. This means that these networks must be able to interwork with the IMS architecture. The procedures already described apply to the IMS network, not the legacy network. The only thing that changes will be the events and procedures implemented at the access layers of the network prior to reaching the IMS. Once the IMS is reached, all of these procedures become valid.

The traditional wireline network typically relies on the control protocol Signaling System #7 (SS7). Like the IMS and SIP, the SS7 protocol is used to establish and manage voice sessions. However, SS7 does not support other media types; hence the need to move to a new architecture and network model such as IMS.

SS7 relies on a separate network for call control, just like the IMS. This network is often referred to as the Intelligent Network (IN). The idea behind the IN was to create a control network that all network switches would connect to, and move the call intelligence and control into the core of the network. This in theory would reduce operations cost and allow operators to implement new services more quickly, without having to enable every switch in the network to support these new services.

There were many obstacles to full implementation of the IN concept, not the least of them cost. But many of the tier 1 network operators have implemented this model and have been relying on the IN architecture for decades. Certainly it is the IN that supports services such as Freephone, Calling Name Display, and Number Portability, all of these being implemented worldwide. Of course, wireless networks could not support roaming today without the use of SS7 and its architecture.

The IN architecture is really quite simple. Each switch in the network is equipped with a service switching point (SSP) function, which allows the switch to generate SS7 messages and interact with the rest of the network via SS7. The switch in turn can be connected to other switches, or connected through a hub and router referred to as the signal transfer point (STP). The STP provides many services outside the scope of this book (my Signaling System #7 book goes into much greater detail of SS7 and the IN), but for the purposes of this topic the STP becomes a focal point in interworking with the IMS.

Applications and services are deployed on servers referred to as service control points (SCPs). These servers are large, dedicated servers that are SS7 specific but will also have to interwork with the IMS if operators want to utilize their legacy services in an IMS environment. This is discussed in more detail in Chapters 1 and 2.

The gateway into the PSTN from the IMS network is the Media Gateway Control Function (MGCF). It is up to the MGCF to convert SS7 messages into SIP messages for forwarding into the IMS network. The MGCF acts as a SIP user agent in the IMS environment, and as an SSP in the SS7 environment.

The MGCF communicates with other switches in the same network using SS7 on the legacy side, but when the MGCF must communicate with switches in another network, it does so through the Breakout Gateway Control Function (BGCF). Think of the BGCF as the gateway into a network, providing a firewall function from one network to another. The MGCF then manages communications within one network domain interfacing between the SS7 and the SIP domains in one network domain. This is illustrated in Figure 5.2.

Figure 5.2 simplifies the relationship between the BGCF and the S-CSCF between networks by eliminating all of the other network elements. Chapter 1 provides additional information about the role of the BGCF and the MGCF.

When the S-CSCF receives an *INVITE* for a subscriber outside of the IMS, the S-CSCF forwards the *INVITE* to the BGCF. It is then up to the BGCF to determine how to route the *INVITE* into the PSTN. If the *INVITE* is addressed to a subscriber within the same network, then the BGCF will route the message to the MGCF for connecting

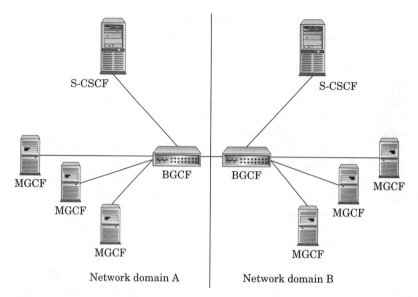

Figure 5.2 Relationship between the BGCF and MGCF

into the SS7 domain. However, if the subscriber is in another network, the BGCF will then forward the message to the BGCF within that network. The S-CSCF uses the TEL URI to determine which BGCF should receive the *INVITE.*

When the MGCF receives an *Initial Address Message (IAM)* from the SS7 network, it will generate an *INVITE* and forward the *INVITE* to the I-CSCF within its own network (if the subscriber is addressed in the same network). Otherwise, the *INVITE* is sent to the BGCF for forwarding to the proper network.

The subscriber's address obviously must be converted from a telephone number into a SIP URI for routing within the IMS, so it is up to the MGCF to convert the telephone number into a TEL URI for routing within the SIP domain. This is done through a query to the ENUM function supporting the network. The ENUM function maps URIs to E.164 numbers, and vice versa.

The MGCF will also include a *SUPPORTED* header set to the value of 100REL (indicating the call originated within the PSTN). The *P-ASSERTED-IDENTITY* header is populated with the subscriber TEL URI, and the *P-CHARGING-VECTOR* header is populated with a new and unique *ICID.* The *ORIG-IOI* parameter is set to the value of the MGCF network.

For calls that originate in the IMS and terminate in the SS7 domain, the MGCF will convert the *INVITE* to an SS7 *IAM,* and will return a response 100 Trying to the IMS originator. The MGCF will then match the codecs at the media gateway (MG) and send the response 183 Session Progression to the originator in the IMS.

Once the SS7 network returns the *Address Complete Message (ACM)* indicating that the subscriber is being notified of an incoming call (the phone is ringing), the MGCF sends a response 180 Ringing to the IMS. When the call is answered and the MGCF

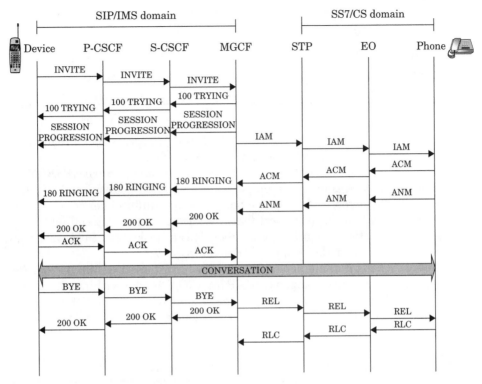

Figure 5.3   Call flow between the SIP and SS7 domains

receives the *Answer Message (ANM)* from SS7, it sends the response 200 OK to the IMS originator. Figure 5.3 shows a typical call flow between the two domains.

There are numerous variables for the responses, depending on the values within the SS7 parameters (if originated in the SS7 domain), or dependent on the SIP header values (if originated in the SIP domain). Tables 5.2 and 5.3 show the SIP responses to the various SS7 messages given different parameter values.

If a *Confusion (CON)* message is received, the MGCF will usually discard the message without sending any SIP response. There is an option of encapsulating the *CON* message into a SIP message body and forwarding it through the IMS to another SS7 entity, but there is no mapping of parameter values to SIP in this case.

When the MGCF receives an *Exit Message (EXM),* the SS7 *EXM* is encapsulated within SIP and sent through the IMS to the destination entity, using the response

TABLE 5.2   SIP Response to an Address Complete (ACM) Message

Backward Call Indicator/Called Party's Status Indicator Parameters	SIP Response
00 No indication	183 Session Progress
01 Subscriber free	180 Ringing

TABLE 5.3   SIP Response to a Call Progress (CPG) Message

Event Information/Event Indicator Parameters	SIP Response
000 0001 Alerting	180 Ringing
000 0010 Progress	183 Session Progress
000 0011 In-band info	183 Session Progress

183 Session Progress. If a response *183 Session Progress* has already been sent, then the *EXM* is encapsulated within an *INFO* message.

There is no mapping for *Pass Along Message (PAM)* or *Suspend (SUS)*. If encapsulation is being used then these messages are simply encapsulated within a SIP *INFO* message. The purpose of these two SS7 messages differ dependent on implementation, but are typically used to support features such as three way calling (the switches hold the call connection when the subscriber does a hook flash to initiate these features). There is no need for these in the SIP domain.

There is also mapping of cause code values contained within an SS7 *Release (REL)*. These cause codes are mapped to a SIP *REASON* header within a SIP *BYE* or *CANCEL*. Likewise, there is mapping between specific cause codes and SIP responses. Tables 5.4 and 5.5 provide the mapping between SS7 cause codes and SIP responses.

TABLE 5.4   Mapping ITU SS7 Cause Codes to SIP Responses

SS7 Cause Code	SIP Response
1 Unallocated number	404 Not Found
2 No route to network	500 Server Internal Error
3 No route to destination	500 Server Internal Error
4 Send special information tone	500 Server Internal Error
8 Preemption	500 Server Internal Error
9 Preemption circuit reserved for re-use	500 Server Internal Error
17 User busy	486 Busy Here
18 No user responding	480 Temporarily Unavailable
19 No answer from the user	480 Temporarily Unavailable
20 Subscriber absent	480 Temporarily Unavailable
21 Call rejected	480 Temporarily Unavailable
22 Number changed	410 Gone
27 Destination out of order	502 Bad Gateway
28 Invalid number format (address incomplete)	484 Address Incomplete
29 Facility rejected	500 Server Internal Error
31 Normal unspecified	480 Temporarily Unavailable

TABLE 5.4   **Mapping ITU SS7 Cause Codes to SIP Responses (*continued*)**

SS7 Cause Code	SIP Response
34 No circuit/channel available	480 Temporarily Unavailable
38 Network out of order	480 Temporarily Unavailable
41 Temporary failure	480 Temporarily Unavailable
42 Switching equipment congestion	480 Temporarily Unavailable
43 Access information discarded	480 Temporarily Unavailable
44 Requested circuit/channel not available	480 Temporarily Unavailable
46 Precedence call blocked	480 Temporarily Unavailable
47 Resource unavailable—unspecified	480 Temporarily Unavailable
50 Requested facility not subscribed	500 Server Internal Error
57 Bearer capability not authorized	500 Server Internal Error
58 Bearer capability not presently available	500 Server Internal Error
63 Service option not available, unspecified	500 Server Internal Error
65 Bearer capability not implemented	500 Server Internal Error
69 Requested facility not implemented	500 Server Internal Error
70 Only restricted digital information bearer capability is available	500 Server Internal Error
79 Service or option not implemented—unspecified	500 Server Internal Error
88 Incompatible destination	500 Server Internal Error
91 Invalid transit network selection	404 Not Found
95 Invalid message	500 Server Internal Error
97 Message type non-existent or not implemented	500 Server Internal Error
99 Info element/parameter non-existent or not implemented	500 Server Internal Error
102 Recovery on timer expiry	480 Temporarily Unavailable
103 Parameter non-existent or not implemented, passed on	500 Server Internal Error
110 Message with unrecognized parameter, discarded	500 Server Internal Error
111 Protocol error, unspecified	500 Server Internal Error
127 Interworking, unspecified	480 Temporarily Unavailable

SIP provides a bit more granularity about failed or terminated sessions. Because of this there are more options for possible SIP responses than there are SS7 cause codes to map. Table 5.6 shows how SIP responses are mapped to SS7 cause codes (from SIP to SS7).

Figure 5.3 is a simplified call flow, but it does illustrate the difference between SIP and SS7 signaling, as well as the relationship between the two protocols. There is a lot more conversation between network entities within the SIP/IMS domain than there is within the SS7 domain.

TABLE 5.5   Mapping ANSI SS7 Cause Codes to SIP Responses

SS7 Cause Code	SIP Response
23 Unallocated number	404 Not Found
25 Exchange routing error	500 Server Internal Error
26 Misrouted call to a ported number	404 Not Found
45 Preemption	500 Server Internal Error
46 Precedence	500 Server Internal Error
51 Call type incompatible with service request	500 Server Internal Error
54 Call blocked due to group restrictions	500 Server Internal Error

TABLE 5.6   Mapping SIP Responses to SS7 Cause Codes

SIP Response	SS7 Cause Code
400 Bad Request	127 Interworking
401 Unauthorized	127 Interworking
402 Payment Required	127 Interworking
403 Forbidden	127 Interworking
404 Not Found	1 Unallocated number
405 Method Not Allowed	127 Interworking
406 Not Acceptable	127 Interworking
407 Proxy Authentication Required	127 Interworking
408 Request Timeout	127 Interworking
410 Gone	22 Number changed
413 Request Entity Too Long	127 Interworking
414 Request-URI Too Long	127 Interworking
415 Unsupported Media Type	127 Interworking
416 Unsupported URI Scheme	127 Interworking
420 Bad Extension	127 Interworking
421 Extension Required	127 Interworking
423 Interval Too Brief	127 Interworking
480 Temporarily Unavailable	20 Subscriber absent
481 Call/Transaction Does Not Exist	127 Interworking
482 Loop Detected	127 Interworking
483 Too Many Hops	127 Interworking
484 Address Incomplete	28 Invalid number format
485 Ambiguous	127 Interworking
486 Busy Here	17 User Busy

TABLE 5.6   Mapping SIP Responses to SS7 Cause Codes (*continued*)

SIP Response	SS7 Cause Code
487 Request Terminated	127 Interworking
488 Not Acceptable Here	127 Interworking
491 Request Pending	No Mapping
493 Undecipherable	127 Interworking
500 Server Internal Error	127 Interworking
501 Not Implemented	127 Interworking
502 Bad Gateway	127 Interworking
503 Services Unavailable	127 Interworking
504 Server Timeout	127 Interworking
505 Version Not Supported	127 Interworking
513 Message Too Large	127 Interworking
580 Precondition Failure	127 Interworking
600 Busy Everywhere	17 User Busy
603 Decline	21 Call Rejected
604 Does Not Exist Anywhere	1 Unallocated number
606 Not Acceptable	127 Interworking

This means that service providers can expect a significant increase in their signaling traffic, and if they are capturing this traffic for analysis and functions such as revenue assurance, they will see a marked increase in storage capacity requirements. This is because within SS7, the network is only concerned with communicating about physical connections. In the SIP network, the protocol is communicating everything about the subscriber, including the subscriber's current location (address).

Messages between the SS7 and SIP domains must be mapped between the two protocols. When a SIP *INVITE* is received at the MGCF, the MGCF must derive certain parameters and headers and use these values to create the equivalent SS7 messages. Following is a description of the SS7 parameters and how they are mapped into SIP.

- Called party number
- Calling party number
- Calling party's category
- Forward call indicators
- Hop counter
- Nature of connection indicators
- Transit network selection
- User service information

The parameters and their respective values are discussed further in the text that follows. The intent is not to provide an exhaustive accounting here, but enough information as to understand the role these parameters play within each domain and how they interact between the legacy network and the IMS.

Called Party Number   This is a mandatory parameter from the SS7 network identifying the digits dialed by the calling party and that is used to derive the *REQUEST-URI* used in the SIP *INVITE*. The calling party telephone number is converted into a TEL URI for the SIP messaging.

Calling Party Number   The calling party number is mapped to the *P-ASSERTED-IDENTITY* header in SIP. From this header the MGCF can determine the calling party from the SIP domain. The calling party number parameter in the SS7 message is part of the *IAM*.

There are two parameters in the *Calling Party Number* field of SS7 that must also be set if being sent back into the SS7 network. The outgoing SS7 message is populated with the *Numbering Plan Indicator* set to *ISDN (telephony) Numbering Plan*. The *Nature of Address Indicator* is also examined to determine if the number is a national or an international number. The MGCF must be able to read the country code (CC) portion of the SIP TEL URI found in the *P-ASSERTED-IDENTITY* to determine this.

Calling Party's Category   This is also a mandatory parameter for the SS7 message. The value is set to 0000 0000 Calling party's category unknown when converting from a SIP *INVITE* to an SS7 *IAM*. If the call is an emergency call, this can be set to NS/EP Call or Emergency Service Call, dependent on the operators' implementation.

Forward Call Indicators   The Forward Call Indicators are found in the IAM and must be set as follows:

Bit D	Interworking indicator	Interworking encountered	1
Bit F	BICC/ISUP indicator	BICC/ISUP used all the way	0
Bits HG	BICC/ISUP preference indicator	BICC/ISUP not required all the way	01
Bit I	ISDN Access indicator	Originating access non-ISDN	0
Bit M	Ported number translation indicator	Number not translated	0
		Number translated	1

Hop Counter   The *Hop Counter* is mapped to the *MAX-FORWARDS* header in the SIP domain. If the call originated in the SS7 domain, the *Hop Counter* value is used for the *MAX-FORWARDS* header, and decremented by one.

In the opposite direction, the *MAX-FORWARDS* header value is copied into the *Hop Counter* parameter of the SS7 message and decremented by one.

Nature of Connection Indicators   There are a number of parameters within the SS7 *IAM* that must be set in the SS7 direction if a SIP *INVITE* is received by the MGCF. These parameters are:

Bits AB	One satellite circuit in the connection	01
Bits DC	Continuity check not required (ISUP) or No COT to be expected (BICC)	00
	Continuity check performed on a previous circuit (ISUP) or COT to be expected (BICC)	10
BIT E	Outgoing echo control device included	1

Transit Network Selection   The *Carrier Identification Code (CIC)* is derived from the *REQUEST-URI* if available and used to identify the transiting carrier. In the SS7 domain, this parameter is set to identify any transit networks that were used to carry the call to its destination. It may or may not be populated.

User Service Information   User Service Information is mapped into the Session Description Protocol (SDP) found in the message body of the SIP message. Likewise, the media, bandwidth, and access lines are mapped from the SIP SDP into the USI parameter in the SS7 message. Tables 5.7 and 5.8 provide mapping between the two protocols.

There are a number of different responses that can be sent on either the SS7 or the SIP sides of the network.

Interworking with VoIP

When interfacing with another SIP-based network, the network access is the same as to the PSTN. In other words, access to the IMS network is granted through the MGCF and BGCF functions.

There is no interworking between protocols. The IMS entities should support simple SIP (based on RFC 3261), so interworking with other SIP entities is not an issue. When sending SIP messaging with IMS-specific extensions into the VoIP network, the VoIP entities respond according to the procedures outlined in RFC 3261.

The S-CSCF is capable of supporting simple SIP as well and is therefore able to process these messages when received from a VoIP network. When forwarding messages to a VoIP network outside of the IMS domain, SIP procedures are used.

Establishing Sessions in the IMS Using SIP

So far we have covered accessing the IMS and registration with the IMS network. All of these processes must be completed prior to a session being established. As mentioned earlier, the IMS uses SIP procedures for session establishment; however, there are a number of SIP extensions that have been defined specifically for the IMS.

It is also worthy of noting that establishing a session within the IMS is quite a bit more robust than in most VoIP networks. For example, it is perfectly acceptable to access

TABLE 5.7   Mapping Between the SIP SDP and SS7 USI/HLC (SIP Origination)

SIP SDP			SS7 User Service Information			High-Layer Characteristics
Media Line (m=)	Bandwidth Line (b=)	Access Line (a=)	Info Transfer rate	Info Transport Capability	User Info Layer 1 Protocol Indicator	High-Layer Characteristics Identification
audio RTP/AVP 0	Nothing or 64 Kbps		64 Kbps	3.1 KHz audio	G.711 μ-law	
audio RTP/AVP	64 Kbps	rtpmap PCMU/8000	64 Kbps	3.1 KHz audio	G.711 μ-law	
audio RTP/AVP 9	AS:64 Kbps	rtpmap: 9 G.722/8000	64 Kbps	Unrestricted digital info with tones and announcements		
audio RTP/AVP	AS: 64 Kbps	Rtpmap CLEARMODE/8000	64 Kbps	Unrestricted digital info.		
image udptl t38	Up to 64 Kbps		64 Kbps	3.1 KHz audio		Facsimile Group 2/3
image tcptl t38	Up to 64Kbps		64 Kbps	3.1 KHz audio		Facsimile Group 2/3
audio RTP/AVP	384 Kbps	rtpmap CLEARMODE/8000	384 Kbps	Unrestricted digital info		
audio RTP/AVP	1472 Kbps	rtpmap CLEARMODE/8000	1472 Kbps	Unrestricted digital info		
Audio RTP/AVP	1536 Kbps	Rtpmap clearmode/8000	1536 KBPS	Unrestricted digital info		

a VoIP network and establish a session without any registration or authentication (possible but not wise). Much has been added in terms of procedures for IMS networks in an effort to make the network more secure and reliable. That is not to suggest that IMS and SIP is perfect, but there have been significant improvements over simple SIP used in VoIP.

Basic Session Establishment

A session can be any form of communication and is not limited to a voice call. This includes (but is not limited to) e-mail, text messaging, instant messaging, and even interactive sessions such as Webinars and gaming sessions. A call may consist of more than one session if other media types are involved in the call. For example, if video is used as part of the call, the voice will comprise one session and the video another.

The CSCFs within each IMS network act as proxies within the IMS domain, and therefore follow the procedures outlined in RFC 3261 for both stateless and stateful proxies. There are some instances when the I-CSCF, for example, may act as a stateful proxy.

TABLE 5.8 Mapping Between SS7 USI and the SIP SDP (SS7 Origination)

SS7 User Service Information				SIP SDP		
Info Transfer Rate	Info Transport Capability	User Info Layer 1 Protocol Indicator	High-Layer Characteristics Identification	Media Line (m=)	Bandwidth Line (b=)	Access Line (a=)
Speech	Speech	G.711 μ-law		audio RTP/AVP 0	AS:64	rtpmap:0 PCMU/8000
Speech	Speech	G.711 μ-law		audio RTP/AVP 0	AS:64	rtpmap: PCMU/8000
3.1 KHz audio	3.1 KHz audio	G.711 μ-law	Telephony (or no HLC)	audio RTP/AVP 0	AS:64	rtpmap:0 PCMU/8000
3.1 KHz audio	3.1 KHz audio		Facsimile Group 2/3	image udptl t38	AS:64	
3.1 KHz audio	3.1 KHz audio		Facsimile Group 2/3	image tcptl t38	AS:64	
64 Kbps unrestricted	Unrestricted digital info with tone announcement			audio RTP/AVP 9	AS:64	rtpmap:9 G.722/8000
64 Kbps unrestricted	Unrestricted digital info			audio RTP/AVP	AS:64	rtpmap: CLEARMODE/8000

The CSCFs determine the routing of a SIP request based on the URI provided by the subscriber device (and contained in the SIP request).

To establish a session, the originating subscriber device must first generate an *INVITE* message with a destination address in the form of either a SIP URI or a TEL URI. If the *INVITE* is to be directed outside of the IMS and into the legacy network, then the destination address will take the form of a TEL URI. The CSCFs will eventually route the request to the BGCF/MGCF for routing into the legacy network. The BGCF/MGCF will determine how to route the *INVITE* into the legacy network according to this address.

The *INVITE* must then be routed through the network to the destination device. The destination device could be another subscriber device, or it may be an application server providing a service to the subscriber (such as a Web server supporting a Webinar session). The next session looks at how the *INVITE* is routed followed by a discussion regarding the establishment of a dialog, and session management.

Routing the Request   The 3GPP has specified the use of loose routing in the IMS (vs. strict routing) in accordance with RFC 3261. However, actual implementation by operators may use strict routing as an alternative. There are many instances outlined in the various 3GPP documents that describe strict routing, so one can assume that while loose routing is supported, the recommendation is for strict routing.

There are many advantages to strict routing, as there are to loose routing. In loose routing the network makes routing decisions based on traffic in the network and the status of the various network elements. Traditional data networks may choose loose routing simply because this is how they route traffic today.

Strict routing, on the contrary, is more controlled and can cause unbalanced traffic in the network. However, traditional voice service providers use this method today and are likely to use strict routing in their IMS implementations for security purposes. The procedures discussed here are per the 3GPP specifications for IMS.

When a subscriber is establishing a session, the subscriber device will initiate an *INVITE* and forward that *INVITE* to the P-CSCF. The P-CSCF is always the first point of access in the IMS network. The P-CSCF provides some protection to the network by maintaining the state of each registered subscriber, and providing access to the IMS only when it is assured the subscriber is legitimate. The P-CSCF first examines the *REQUEST-URI* to determine if the calling party is in the same network as itself. If both are in the same network, the P-CSCF will then determine the S-CSCF assigned to the calling subscriber and forward the *INVITE* to that S-CSCF. The P-CSCF will know the address of the assigned S-CSCF because it saves this information during the registration process. The P-CSCF will save this information for the entire duration the subscriber is registered.

The *REQUEST-URI* contained in the *INVITE* is critical to routing the *INVITE* outside the network as well. The P-CSCF uses the address in the *REQUEST-URI* (specifically the domain name) to determine how the *INVITE* should be routed in the event the calling party is assigned to an S-CSCF in another network. The domain name will identify the home network of the calling subscriber, and the location of the S-CSCF assigned to the calling subscriber.

The *INVITE* is not sent directly to the S-CSCF in another network. The domain name in the *REQUEST-URI* must be resolved to determine the IP address of the I-CSCF providing access into the destination network. The I-CSCF must then determine how to route the *INVITE* to the proper S-CSCF.

However, the *REQUEST-URI* does not necessarily provide the identity of the subscriber's public user identity (depending on the method of routing used). The P-CSCF relies on the *P-PREFERRED-IDENTITY* header to identify the public user identity to be used for the session request. If there is no header, the P-CSCF then uses the first *ROUTE* header, which should be populated with the default public user identity for the session.

The P-CSCF must then validate the request by comparing the *RECORD-ROUTE* headers with the route list that the P-CSCF created when the subscriber registered with the network. Each individual entry is examined individually to determine if the address listed in the message is part of a legitimate route. If the two do not match, it is possible that the message is coming from another source other than the subscriber, and the request is therefore rejected. The P-CSCF will send a 400 Bad Request response to the originator.

This is yet another security feature provided by the P-CSCF. The P-CSCF can route the message based on the route headers stored from the registration, which would result in the message being routed to the proper subscriber, but ultimately being rejected by the subscriber device because the message did not come from the device (it is fraudulent).

If the message is legitimate, the message is forwarded on. The P-CSCF can then add its own address in the form of domain name and IP address to the *VIA* header, and its SIP URI to the *RECORD-ROUTE* header. The P-CSCF will also delete the *P-PREFERRED-IDENTITY* header and replace it with the *P-ASSERTED-IDENTITY* header populated with the public user identity of the requestor. The *P-CHARGING-VECTOR* populated with the *ICID* is also added to the message before forwarding the message on to the I-CSCF for routing to the proper S-CSCF.

By using the domain of the called party, the P-CSCF can locate the address of the I-CSCF for the network domain being called; it then forwards the *INVITE* to the I-CSCF. The P-CSCF does not know any of the addresses within other networks, only the addresses of the I-CSCF for other networks.

There are several ways to address the I-CSCF. Some operators may choose to assign a common SIP URI to all of their I-CSCF entities (to facilitate load sharing, for example), while others may choose to implement unique addressing. Each I-CSCF will also be assigned its own unique IP address, which is what the SIP URI gets resolved to.

When topology hiding is implemented, outside networks will only know the address of the I-CSCF (or multiple I-CSCFs, as will be the case in most networks). The I-CSCF must then determine how to route requests and responses within its own network.

When the I-CSCF receives the *INVITE,* it must first query the HSS to determine which S-CSCF has been assigned to the called party. The I-CSCF will also store the value in the *P-CHARGING-VECTOR* header, and if there is no *IMS CHARGING ID (ICID)* present, create a new one.

If the I-CSCF queries the HSS and the SLF, but cannot locate the subscriber, the I-CSCF will return the response 404 Not Found to the request originator. If the subscriber is not a user of the home network, the I-CSCF will send the response 604 Does Not Exist Anymore. If the subscriber is a valid subscriber in the network but is not currently registered, the I-CSCF will respond with 480 Temporarily Unavailable.

The *INVITE* is then forwarded to the assigned S-CSCF. The S-CSCF then determines the location of the called party. Prior to forwarding the message to the called party, the S-CSCF will add the *P-CHARGING-FUNCTION-ADDRESSES* header to the request (as long as the request is being forwarded within the S-CSCF network). Charging headers and parameters are not typically sent to outside networks by any entity unless it is a trusted network. The request is then forwarded to the device.

Not all requests require strict routing and the use of the *RECORD-ROUTE* headers. Some services, for example, may bypass the S-CSCF entirely, such as Presence. The general rule of thumb cited by 3GPP specifications is that any services requiring media control, CDR generation, and possibly network-initiated session termination require strict routing to be used (and subsequently the use of *RECORD-ROUTE* headers).

Media Negotiation and Establishing the Session   So far we have looked at how a request is routed through the network, and how the responses are sent in return. Now let's look at the procedures at both the origination and the destination devices for establishing a session. To maintain clarity, we will repeat some of the previous procedures.

We have already learned that to establish a session with another device, the originating subscriber device must first initiate a request (usually in the form of an *INVITE* message). This is the first step in establishing a dialog with the destination device. This dialog will be maintained throughout the session and allows both devices to communicate with one another regarding the established session.

To establish the dialog, the originating device must first receive a final response from the destination device. A final response in this case must be a successful response (2*xx* class of responses). If any other final response is received, there was an error and the session cannot be established.

The dialog is not the same as a session. The dialog is established between two entities and requires a specific sequence of events to take place prior to the dialog being established. This sequence of events is the acceptance of a request, and successful responses to that request.

When the originator receives a 200 OK, for example, the device will return an *ACK*, establishing the dialog between the two devices. The dialog is identified by a combination of headers and parameters, and each device uses the values of these headers and parameters to calculate a dialog ID. Only *INVITE, SUBSCRIBE,* and *REFER* can be used to create a dialog between two devices (note that the device may be an application server and is not always a subscriber device).

The dialog ID is calculated by using the *TAG* value from the *TO* and *FROM* headers, and the *CALL ID*. When the request is created, the originator will calculate its own *tag* value and append it to the *FROM* header. The destination device will calculate its own *tag* value and attach it to the *TO* header upon responding. The 200 OK response carries both tags and of course the *CALL ID* that was sent in the request.

The *tag* values are used by each of the receiving devices to correlate requests with responses throughout the life of the session (and the associated dialog). The dialog ID is used to correlate requests and responses between two devices. They are separate, since a device could establish a session with multiple devices, all containing the same *CALL ID*. The originator will correlate responses to sessions using the *CALL ID*, while correlating responses from each device using the dialog ID (the only difference is the tag values sent by each device).

It is this nuance that differentiates a dialog ID from the *CALL ID*. The dialog is a specific set of communications between two unique devices regarding the specified session. The *CALL ID* identifies the session itself, and differentiates one session from another for all devices involved. The dialog ID is only known by two entities communicating with one another and is not shared by more than two devices.

Now that you understand how a dialog is established, let's discuss resource negotiation. Resource negotiation is the first part of the dialog, where the destination device examines the received request and determines whether or not it will be able to support the media type and other requirements for the requested session.

If the device is able to support the session, the destination device must agree to the session. This is done through the use of a final response. The destination device looks at the SDP in the message body to determine if the requested session can be supported by the device. This is the first phase of session negotiation. If the device cannot support the requested session, the device will return a response of 488 Not Acceptable Here and reject the request.

When the *INVITE* is received by the called subscriber, there are two options. The first option is for the device to process the *INVITE* and send the appropriate response, given the requirements set forth in the *INVITE* SDP (codecs required, media type supported, etc.). This is the usual method.

The other option is for the device to alert the user with some form of message and allow that user to accept or deny the session in accordance with the media type required. For example, if the session requires a text message to be received by the subscriber, the operator may want to give the subscriber the option of approving the session prior to establishing it. This last option is beyond the scope of the IMS specifications but is being reviewed by the industry as a possible option.

If the device is able to support the session, the device will send the response 200 OK using the same path that was used to route the request. At this point the destination subscriber device has calculated the dialog ID for the session and has saved this dialog ID for subsequent requests and responses for the same session.

Once the requestor has received the response from the called subscriber, an *ACK* can be returned and the session can begin. The *ACK* contains the same *BRANCH* value as found in the *VIA* header in the request. A sample of these messages follows.

```
INVITE sip:travis.russell@tcg.com sip/2.0
VIA: SIP/2.0/UDP  pchome101@aol.com:5060;branch= z9hG4bK74gh5
FROM: Deby Russell <sip:deby.russell@aol.com>; tag=9hz34567sl
TO: Travis Russell travis.russell@tcg.com
MAX FORWARDS: 70
CALL-ID: 82167534@126.18.27.0
CSeq: 1 INVITE
```

```
CONTACT: Deby Russell <sip:deby@126.18.27.0>
CONTENT-TYPE: application/SDP
CONTENT-LENGTH: 154
```

This *INVITE* contains the basic headers needed for session establishment. In an IMS environment there are numerous other headers, but they have been eliminated in this example for simplicity.

Notice that the *TAG* value has been calculated in the *FROM* header, but not yet the *TO* header. This is because the destination must calculate its own unique *TAG* value and append it to the response. Here is an example of a 200 OK response:

```
SIP/2.0 200 OK
VIA: SIP/2.0/UDP pchome101@aol.com:5060; branch=z9hG4bK74gh5
FROM: Deby Russell <sip:deby.russell@aol.com>; tag=9hz34567sl
TO: Travis Russell <sip:travis.russell@tcg.com>; tag=1df789jkf
MAX FORWARDS: 70
CALL-ID: 82167534@126.18.27.0
CSeq: 1 INVITE
CONTACT: Travis Russell <sip:travis.russell@135.18.10.10>
CONTENT-TYPE: application/SDP
CONTENT-LENGTH: 154
```

In this example, the *TAG* value has now been entered into the *TO* header, and the request originator can now calculate the dialog ID. The parameters shown in bold are used for calculating the dialog ID.

The request originator will now generate the *ACK* in response to the *200 OK* and the session is now established. Notice in the *ACK* that the *BRANCH* value in the *VIA* header is the same as in the *INVITE*. This is how stateful proxies within a SIP environment correlate requests and responses (for example, any CSCF acting as a stateful proxy).

```
ACK sip:travis@135.18.10.10 SIP/2.0
VIA: SIP/2.0/UDP pchome101@aol.com:5060; branch=z9hG4bK74gh5
FROM: Deby Russell <sip:deby.russell@aol.com>; tag=9hz34567sl
TO: Travis Russell <sip:travis.russell@tcg.com>; tag=1df789jkf
MAX FORWARDS: 70
CALL-ID: 82167534@126.18.27.0
CSeq: 1 INVITE
CONTACT: Deby Russell <sip:deby.russell@128.18.27.0>
```

One other note regarding the *BRANCH* value: the first seven digits are always z9hG4bK (indicating this came from SIP 2.0 in accordance with RFC 3261).

There is of course much more to the entire process that is outside the scope of this book. You can refer to the RFC 3261 for detailed specifications for SIP processing and session establishment, or refer to my other book, *"Session Initiation Protocol (SIP): Controlling Convergent Networks* (McGraw-Hill, 2008).

Emergency Session Establishment

The P-CSCF is responsible for emergency session establishment. The P-CSCF will store and maintain an operator configurable list of local (local to the P-CSCF and within its domain) emergency numbers and URIs. The P-CSCF can also maintain a list of emergency numbers and URIs for any roaming and interconnect partners.

When a session request (i.e., *INVITE*) is received by the P-CSCF containing the TEL or SIP URI of an emergency number (such as 911), the P-CSCF will route this request directly to the appropriate destination.

The P-CSCF will send the response 380 Alternative Service to the originator of the request, indicating that there will not be a dialog established, and no 2*xx* response will be sent. The request may be appended by the P-CSCF prior to forwarding to the destination to indicate this is an emergency call.

There is still work underway to define how some government service will be supported, such as the Government Emergency Telephone System (GETS) in the U.S. The 3GPP as well as the NS/EP IR working group are addressing the needs of government for the prioritization of emergency calls (such as precedence in military applications) and how these will be implemented in SIP and the IMS. This work will likely continue over the next couple of years.

Modifying SIP Sessions in the IMS

Any established session can be modified while the session is in progress. For example, during a conference call the originator may wish to add video into the call. This is achieved by sending a new request to each participant. Any participant in an established session can modify the session (dependent on applications, of course).

Note, however, that not all participants must receive this new request. If the originator only wishes some of the participants to be part of the video session, the request is sent only to those participants.

When the new request is sent, it will be populated with the same *CALL ID* as the session in progress. This alerts devices that this is not a brand new session but a modification to an existing session. The new request will identify the participants and the media to be added, and any other modifications to be made.

The SDP in the message body will contain a description of the entire session, as well as any changes being made to the session. This is to keep all devices in synch and prevent any confusion between devices.

The SDP will also contain a version field to indicate a change is being made to the original session. For example, the original session would be version=1, while a change to that session would be version=2. The entire session description is checked for any changes from the original *INVITE*.

Typically the *INVITE* method is used to modify a session. However, this is not always the case. The general rule of thumb is to use the *INVITE* method if there is to be some form of user interaction (such as acceptance from the user prior to starting a video session). Otherwise, the *UPDATE* method can be used.

All the procedures discussed earlier regarding routing and session establishment apply to session modification as well. Devices must still accept the request by responding with a successful response; otherwise, the session modification is rejected. The original session is not usually interrupted if the modification is rejected.

Terminating SIP Sessions in the IMS

Any participant in a session can terminate the session. This is done by using the *BYE* method to the other participants in the session. Each participant will then respond with a 200 OK. There are some instances when the network itself wishes to terminate a session.

For example, loss of signal to a device would leave a session in suspension. The physical connection is lost, but any sessions would remain until they either timed out or the network terminated them.

There are other cases where the network must terminate a session in progress. The S-CSCF acts as a stateful proxy in the IMS and is therefore capable of terminating session. If the S-CSCF must terminate a session, it will send a *BYE* message in two directions: one to the originator and one to the called party. The *BYE* sent to the called party would be populated with:

- *REQUEST-URI* populated with the address of the called party
- *TO* header populated with the address of the called party derived from the 200 OK
- *FROM* header populated with the address of the calling party derived from the *INVITE*
- *CALL-ID* populated with the value from the *INVITE*
- *CSeq* header incremented by one
- *ROUTE* header populated with the address of the called party
- Any additional headers deemed necessary by the network or operator

The same would be sent to the calling party, with the following headers and parameters:

- *REQUEST-URI* populated with the address of the calling party
- *TO* header populated with the address of the calling party derived from the 200 OK
- *FROM* header populated with the address of the called party derived from the *INVITE*
- *CALL-ID* populated with the value from the *INVITE*
- *CSeq* header incremented by one
- *ROUTE* header populated with the address of the calling party
- Any additional headers deemed necessary by the network or operator

The S-CSCF will expect to receive a 2*xx* response from both parties. The S-CSCF can also cancel a session prior to it being established. If a request has been sent but the *ACK* has not yet been received, the S-CSCF can use the *CANCEL* method to stop the session from being established. The *BYE* method cannot be used because the session has not yet been established.

Security Procedures in the IMS

The telecommunications industry has changed from one of trusted entities to open competition and consequently, a less secure environment. While fraud and security have always been a concern in this industry, never before has there been more opportunity for breaching networks than today with the advent of IP networks. Add the open Internet to the mix, and you have a security officer's nightmare.

Prior to the privatization of telecom networks (and the divestiture of the Bell System in the U.S.), accessing a telephone network was granted only to the very monopolies that managed those networks. Providing access in any fashion was strictly forbidden, and access was heavily protected. Learning about the various nuances of these networks was only possible through specialized training learned on the job working for a telecommunications provider.

Today, almost every major university has degree programs on telecommunications management, teaching courses on the technologies that enable today's communications networks to work. Even subjects such as Signaling System #7 (SS7) can be learned through thousands of open sources today.

Gaining access to ones network only requires an operator's license, which today is easy to obtain. In some countries, all that is required is a commercial bank account, which can be obtained with nothing more than a driver's license. There are many rogue operators dumping traffic on legitimate carriers all over the world; they are highly trained, well funded, and well educated.

The Internet changed the paradigm of communications for everyone, not just in the sense of mail and instant messaging, but now voice communications. Once it became possible to route voice communications over the public Internet, these very same rogue operators found a path to routing calls all over the world for few to no termination fees on their part. Networks continue to be compromised on a global basis, thanks to VoIP.

One of the pitfalls of VoIP is a lack of security. The standards really do not address security implementations and do not provide mechanisms as part of the registration process, leaving this up to the operators' own implementation. Unfortunately many operators never implement any fashion of security, treating the VoIP nodes in the same fashion as their legacy voice switches (which have no security at all from a subscriber perspective). This author has even heard from some operators that have disabled authentication in their networks because they are afraid it will slow down the call processing.

There have been repeated cases of VoIP breaches where hackers have compromised media gateway controllers (MGCs) through open ports, with no authentication or challenge from the MGC. They have then used these open ports to route their own traffic, and have even reconfigured the MGCs to accept traffic from other networks.

Even data servers supporting commercial and public intranets have been compromised and used for the purpose of routing VoIP calls through those networks and into unsuspecting "trusted" telecommunications networks.

The fact is the telephone network was once a "trusted" domain (which is explained a little further down). All of the operators connecting with one another were either from the same big company or a government entity if outside the U.S. The privatization of telephone companies around the world, and the divestiture of the Bell System in the U.S., changed this forever.

When the open and free Internet is suddenly connected to the "trusted" networks of telephone companies, exploitations and breaches become prolific. Without a means of identifying those who access the network, and verifying those identities through the exchange of some fashion of authentication keys, there is no way to stop fraudulent traffic from rogue service providers.

Suddenly anyone can become a telephone company with a little money and licensing from the appropriate authorities. Today there are many operators connecting into this "trusted" domain. While most are legitimate, there are a surprisingly large number of fraudulent or unethical operators who cannot be trusted. While their activities may not be illegal, they are not fair business practices, and they result in the legitimate operators losing precious revenues.

Likewise, there are many organizations using the world's telephone networks for their own gain. They accomplish this by discovering those networks that have easy access with no authentication required. They "hack" into network platforms where passwords have not been changed from their factory defaults. They explore every opportunity to gain unauthorized access into networks and then use this access to sell their own services.

Security Threats in an IP Domain

There are a number of security threats already established for IP networks (and well exploited). These attacks can be prevented through the practices discussed in this chapter and do not necessarily require huge investments to prevent. The main threats today are

- Eavesdropping
- Registration hijacking
- Server impersonation

- Message body tampering
- Session tear-down
- Denial-of-service
- Amplification

Eavesdropping

Eavesdropping lets hackers intercept SIP messages without detection. This method is used by hackers to obtain sensitive information such as routing and private user identities. The information is then used to create new messages that will pass authentication unless there are other forms of authentication and security (as we will continue to discuss).

The eavesdropping is prevented through encryption, both within the secure home domain as well as across transited networks. Encryption is the best means to prevent eavesdropping from occurring and is one of the reasons that IPv6 with IPsec has been defined for use within the IMS.

The wireless network also uses encryption to prevent eavesdropping at the air interface. This was a problem in earlier wireless networks and the driver to implementing encryption across the air interface in GSM and CDMA.

It is just as applicable in wireline networks, where hackers can compromise ports on communications servers to enable copying of messages and rerouting of SIP messages to their own servers. Encryption then becomes useful for preventing this in the wireline networks as well.

Registration Hijacking

Registration hijacking occurs when a *REGISTER* message is sent by a hacker with a subscriber's stolen identity. The identities are learned through eavesdropping and serve to move a current registration to another location or establish an entirely new registration if the legitimate subscriber is not connected and registered. Once registered, the attacker has full access to the same services as the legitimate subscriber.

If a hacker gains access to a subscriber's private user identity, the hacker will be able to launch a *REGISTER* message from anyplace in the network. Remember that a *REGISTER* identifies the subscriber's current location. Therefore, if the subscriber is already registered, and the network receives a new registration, the network registrar (the S-CSCF in IMS) will assume that the subscriber has changed locations, and change the registration to the new location.

All new traffic will then be sent to the newly registered location. The legitimate subscriber has no idea that his or her registration has been hijacked, and simply stops receiving calls. If subscribers themselves change locations or power off their devices, the registration will be changed again when they power their devices back up, and they will restore control over their service. Of course by then the perpetrators have already gained access to the services they were after, and have taken advantage of the free service.

Server Impersonation

Server impersonation allows a proxy server in another network to "pose" or masquerade as another legitimate proxy redirecting all traffic to itself and away from the legitimate proxy. In the IMS domain, this would mean compromising the CSCF function and redirecting requests and responses to the rogue CSCF.

We see this form of attack in the Internet domain already. Web sites that are masquerading as legitimate Web sites are used to coax sensitive information from unsuspecting consumers. This could become a major problem within the IMS if there is no means to verify the network.

Procedures are defined for IMS that allow the user device to authenticate the IMS network to ensure it is communicating with the trusted domain of the service provider (or authorized partner). There are other measures we will talk about later that help prevent this as well.

Message Body Tampering

Tampering with the message body requires access to the message body, which without encryption is very easy. Since SIP is sent in plain text (rather than binary coded), it is very easy to eavesdrop on a SIP network and read the contents of the messages.

Depending on the type of SIP message, this has varying consequences. If the message body contains a text message containing the message "call me" and provides a telephone number, the text message could be modified and delivered to the destination with a different number. The receiver would then respond back to the rogue subscriber rather than the legitimate sender of the message. There are many other examples of how this method could be used to steal identities and obtain personal information from unsuspecting subscribers.

Altering the message body could also result in hijacking of subscriber accounts and services. By capturing an *INVITE,* for example, the headers providing routing information could be captured to then reroute traffic to the perpetrator.

Session Teardown

Tearing down sessions is a disruptive attack that if launched to a wide range of subscribers could have serious implications. For example, if this were carried out in concert with a major catastrophic event, communications for thousands of people would suddenly be compromised, as they would have been "cut off" and attempting to re-connect would congest the network.

There are two reasons why an attacker would use this method. One would be to create a Denial of Service (DoS) attack on a network or network segment. Disconnecting all calls in progress followed by another form of physical attack would leave citizens in a panic, creating just the type of reaction needed to instill chaos.

The other reason would be to simply congest the network to where service was degraded. This is accomplished when everyone that was disconnected suddenly attempts to reconnect, all at the same time. The network is usually unable to handle such large demand and would begin denying service to many subscribers.

Denial-of-Service Attacks

Denial-of-service attacks take many forms and of course can also be very damaging. These types of attacks flood the network to the point where legitimate subscribers can no longer gain access to the communications services they need, eventually leading to the shutdown of many systems when they reach their capacity for traffic.

One way of accomplishing this today is to send a request with a false or spoofed IP address and corresponding *VIA* header making it look like it came from a legitimate subscription. Sending the request to many different entities within the network would result in a flood of responses network-wide.

For example, in an IMS domain, the CSCFs would receive the message, recognize that the IP address was invalid, and respond with a 4*xx* error response. If the message were routed to many different portions of the network, this would result in numerous CSCF functions receiving and responding to the message, causing the congestion in responses.

Of course a DoS attack could just as easily be something as simple as launching millions of session requests at one time, using a call generator or computer. The end result would of course be instant congestion in the network, and the denial of service to all devices attempting connection.

This is actually much easier than most would like to believe. Through the use of BOTs and a BOTNET, millions of calls could be generated into the network simultaneously under one command, resulting in a widespread DoS attack on the entire network.

Amplification

Amplification is similar to denial of service with much broader coverage. The same request is sent to a redirect proxy, which then "splits" the request to many different directions, "amplifying" the number of responses in the network. Instead of having a response coming from one CSCF, the message is split and sent to many CSCFs, all sending responses of 4*xx*.

All of these security threats can be addressed through IMS and should be taken seriously by any IP-based service provider. It should be noted here that these attacks are not unique to IP-based services, though. We see many similar forms of attacks and security threats in the legacy telephone networks around the world today.

Denial-of-service attacks have already been played out in legacy networks, using call generators and other forms of call generation, resulting in the complete blockage of all services for an extended period of time (in some cases hours).

Impersonating service providers, "message body tampering" (in this case the message body is the SS7 signaling message exchanged between two carriers), even session tear-down can be emulated in the existing telephone networks today and many cases have already been documented to demonstrate these as issues.

There are already numerous cases of network providers gaining access to other operators' networks and sending their traffic through these networks. This is why the IMS must be well protected and all possible security measures implemented. From the access level to the signaling plane, security should be a primary concern to all operators.

In fact, if ever there was an argument as to why IMS makes sense, security is that argument. There are a number of functions that have been defined for the IMS to prevent these types of attacks from taking place.

The industry has learned many mistakes with VoIP and has taken to heart the threats that an open Internet model brings to a structured telecommunications environment. This is why the 3GPP set out defining the IMS—partly due to issues with interoperability of VoIP elements, and partly because of the lack of security in early SIP implementations.

In fact, if one looks at the documented cases to date of VoIP security breaches, they involved some form of the attacks already described. Simple SIP as defined by the IETF failed to address many security issues, remaining an open and vulnerable protocol.

The extensions defined by the 3GPP for SIP in the IMS domain address these and many more issues, as we will discuss in this chapter. This is why security is one of the best arguments for IMS, even though there is still some work to be done.

Securing the IMS

There are numerous ways one can secure networks. In GSM networks, subscriber authentication is already implemented. The use of SIMs containing cipher and authentication keys within a GSM phone allows networks to verify the device is legitimate and the subscriber is authenticated.

To maintain security between the user device and the home network, each device is equipped with a Universal Integrated Circuit Card (UICC) that runs the security application. The (UICC) is the little card that fits within the GSM phone. Running on the UICC is the Subscriber Identity Module (SIM) application (which is often what we call the UICC).

The private user identity is kept on the removable SIM, which is placed into whatever phone the subscriber uses. This allows subscribers to move from device to device while taking their identity with them. However, the private user identity is not disclosed. It is kept on the SIM application hidden from users.

The communications between the subscriber module (SIM) and the home network is encrypted and kept secure to prevent another network (or another subscriber) from accessing the private user identity of a subscriber. All of these security functions are kept transparent from the user, since there is no need for the subscriber to know the private identity or anything else kept secure by the operator. This is all transparent and unseen by the subscriber.

Encryption is supported as well over the air interface. This prevents eavesdropping over the radio waves, which is a major concern within GSM circles. There are already a number of devices that support "sniffing" the airwaves and capturing GSM signaling, unless they are encrypted.

Keeping the subscriber confidential is also supported in the GSM world. The concept of the private user identity and the public user identity actually comes from GSM. The private user identity is maintained closely and kept from other networks, so only the home network knows this identity.

Using GSM as a lesson learned, the authors of the IMS and SIP standards have defined six security aspects and security threats associated with each of them:

- Authentication & Authorization
- Confidentiality
  - Eavesdropping
  - Masquerading
  - Traffic Analysis
  - Browsing
  - Leakage
- Denial of Service
- Integrity
- Privacy
- Non-repudiation

Authentication & Authorization

Authentication verifies subscriber devices on the basis of criteria assigned by the operator. Each subscription (and its associated private user identities) is given authentication keys for this purpose. The authentication keys are calculated using an algorithm known to the operator network and the device. This means that the authentication key changes based on the algorithm, making it difficult if not impossible to spoof.

The authentication process is a forced process, rather than implementation specific. The problem with many VoIP implementations today is that authentication is optional, and implementation specific. This results in many service providers not challenging subscribers' devices when they access the network for credentials. Of course, since many operators never implement authentication in their networks as policy, there never are any credentials to share.

The concept of credentials is simple. The operator establishes a set of keys that are used to create credentials that are known only to the operator. These keys are embedded into the subscriber device purchased by the subscriber. To support allowing subscribers to purchase their equipment from anyplace, the credentials are stored on a small Universal Integrated Circuit Card (UICC) as an application. This UICC is then inserted into the subscriber device.

The UICC is another GSM concept that has worked very well for wireless operators. In the U.S., this concept is not implemented as well, because operators want control over which devices their subscribers use in their networks. In other words, they only want devices they sell to be used in their networks. For this reason, the UICC and SIM applications on those UICCs are locked, preventing subscribers from inserting them into other phones and using them.

In networks outside of the U.S., the UICC can be inserted into any device and used on the network. This allows subscribers to purchase their phones from any source and use the UICC to obtain access to the network of choice. This will be an important concept in the IMS, as subscribers will be using a multitude of devices from many different sources and will expect to be able to use these devices on any network they pay to have access on.

This concept is what drives services of many different forms, because subscribers are much more likely to purchase different devices from sources other than their home network provider and use these devices to listen to music or play online games. Operators will need to embrace the concept of the UICC and SIM application if they want their services to be popular and widely used.

On the UICC is an application referred to as the IMS Subscriber Module (ISIM). This is like the SIM card used in GSM phones today. Actually, the little circuit card itself is the UICC, while the application residing on the card is the SIM, but the industry commonly refers to the module as the SIM card.

Also stored on the ISIM is the subscriber's private user identity, also known only to the network. When the subscriber device accesses the network and registers its location, the S-CSCF and the HSS work together to "challenge" the subscriber device. This challenge forces the device to return another registration with the correct credentials based on data provided within the challenge itself (the 401 Unauthorized response is the challenge), and the authentication keys embedded within the ISIM in the device. The following information is stored in the ISIM:

- Private user identity
- At least one public user identity
- Home network domain
- Authentication key
- Ciphering algorithm
- Sequence number checking (SQN)

Authentication also applies for the subscriber. The same authentication keys that are shared between the network and the subscriber device allow the device to authenticate the network when a request is sent to the device (or even a response). This provides another level of assurance that is not available today, preventing subscribers from responding to requests sent by rogue networks.

Authorization   Authorization determines what services a subscriber is allowed access to, as well as what networks the subscriber is allowed to visit. This is stored in the HSS assigned to each subscriber. The service authorization is part of the subscriber's service profile. This means that a subscriber who has multiple public identities may have several service profiles (one for each of his or her public identities), entitling that subscriber to different types of services.

Remember when we talked about identities that a subscriber may have one identity for their PDA, and another identity for their cell phone. Each of these may have a different set of permissions depending on their usage. My work cell phone, as an example, can be used for instant messaging, e-mail, and voice calls. My personal cell phone subscribes to voice, data, picture services, instant messaging, and location-based services (LBS).

Visited networks also access this information when authenticating a subscriber. This is how a visited network determines whether a subscriber is allowed to access the network and use its services. This is another reason for IMS; allowing other networks to share information about a subscription and verify whether or not the subscriber is allowed access to these services.

One of the pitfalls of today's telecommunications networks is the inability to use services while roaming. For example, if I use my cell phone in another country, I cannot use the data service, nor can I use the LBS service. This is because the service providers in the areas I visit do not have the platforms to support these services, and if they did, they would not be compatible with my device.

The concept in the IMS is to allow these other networks to access the application servers in my home network so that they can support the same services I enjoy at home in their networks. My home network provider is then able to earn additional revenues by allowing access to these services to other "trusted" domains, extending my service to any areas I visit.

This concept also came from the wireless world, specifically GSM, where roaming information is shared between roaming partners. It should be noted here, however, that even though the two networks are sharing information about subscribers' ability to roam, the visited network does not have carte blanche access to the home network's HSS.

Authorization prevents subscribers from accessing services they are not entitled to use. When it is combined with authentication and integrity, an operator can be assured that only subscribers authorized to access the network and authorized to use services requested are given service.

Confidentiality

Confidentiality uses encryption to block unauthorized sources from viewing SIP messages. This can be an important aspect of the IMS network where interconnection to other networks is provided. Keep in mind that unlike bit-oriented protocols (such as SS7 and ATM), the SIP messages are in plain text. This means that anyone capable of intercepting the messages, or eavesdropping on the network, will be able to read the full contents of SIP signaling.

This includes authentication data and route lists. If these headers are not encrypted and protected from unauthorized eyes, the operator risks a man-in-the-middle attack and possible session/registration hijacking. If the SIP message contains a text message, the perpetrator is able to read the text message and even intercept it.

Encryption then becomes an important aspect of the IP implementation. This is one of the reasons IPv6 has been defined as critical to IMS implementations. IPv6 is used within a trusted domain for encrypting messages within the network and to prevent eavesdropping within the network. Specifically, IPsec is used within the trusted domain to protect sensitive data from being intercepted within the home network. TLS is used between networks but can be substituted by other methods (these are the 3GPP recommendations).

Another form of confidentiality breach is acquiring the traffic from the network and analyzing the traffic, calculating the time, rate, and length of the session or conversation, the originator of the session, and the destination. This information can then be used to determine a user's location, or if there is an important business decision about to be made.

Traffic analysis can produce a lot of information if the perpetrator has access to the signaling data. The software is readily available to make these calculations, but it should not be assumed that this would always be an external attack. It could be a breach from within the organization if a rogue employee has access to network monitoring equipment.

Of course, there is always sensitive data that is passed between the originating and terminating parties, and between the networks themselves. Payment information, PIN numbers, and other sensitive data can easily be captured in SIP domains and used by rogue employees to gather personal data about subscribers.

This is not a new threat, and we have seen many cases of this in current-day networks. For example, it is not uncommon for calling card PIN numbers to be captured through network monitoring and sold on the black market. This is one reason why internal security and access to the network should be highly protected.

Denial of Service

Network providers today are beginning to understand how serious this threat can be. There have already been numerous documented cases of denial of service attacks made on the legacy PSTN. New technologies always leave openings for exploitation and vulnerabilities not yet discovered, and so denial of service should be considered heavily.

Most of the DoS attacks in today's networks have been focused on a specific target. For example, there are several cases where attackers have used cell phones and hacked conference bridges to call emergency services (such as 911 and 0911). By using a conference bridge, they are able to add an additional call to the bridge, in these cases all of those calls being to the same emergency service. They continue adding calls to the bridge until all of the trunks to the emergency services are blocked.

Similar scenarios have been played out to other targets, using similar techniques. Mass calling events have succeeded in putting switches and trunk groups into congestion, blocking calls for more than an hour in many areas.

Consider the ability to use a computer to generate millions of calls into the VoIP network, and one can see where this could be done quite easily with little to no sophistication. Add the use of BOTs and a BOTNET and the threat becomes many more times serious.

There are many ways that a DoS attack can be launched against a communications network. Of course, simply generating millions of calls in a short time period will usually do the trick and is not all that hard to accomplish (as we've discussed).

With IP this may become a bit more difficult, depending on the architecture of the network, simply because IP increases the bandwidth available (in most cases anyway), but it remains a very high possibility. Call generation can be much easier in an IP environment, and when mixed with other network transactions, easier still. There are other ways to cause a DoS attack in an IP domain without generating calls.

For example, within the IMS domain, simply generating a large volume of registration messages can create a DoS attack on the CSCF resources within a network. This would render the affected portions of the network inaccessible for a given period of time. Text messaging also presents a unique issue that should be taken into consideration. Of course, the generation of text messages is very easy, and since SIP is the carrier for these messages, it presents a unique challenge to the operator.

Already today there are many issues with SMS flooding the networks and causing service outages due to congestion. Most operators have begun offloading their SMS traffic to a specialized packet network so that flooding does not impact the rest of the network. This is because in traditional networks SMS is carried through the SS7 signaling network, and congesting the SS7 network prevents calls from being connected.

Events such as *American Idol* in the U.S. and *Pop Idol* in the UK have created huge amounts of bursty SMS traffic that easily congests the messaging network, and operators have taken measures to engineer the network with this in mind. However, in the IMS the messaging is carried by SIP through the same network used to connect sessions.

It may be worthwhile to investigate an offload process for messaging to get this traffic off of the SIP network onto a specialized network, much like what is done today to offload SMS from the SS7 network (and onto SMPP, a dedicated messaging network).

Another form of attack may be by sending a large volume of *REGISTER* messages into a network, containing the incorrect subscriber credentials. This would overload the S-CSCF and force a number of error responses to be sent in response, not to mention that any device that was already registered could be de-registered. When this occurs, the device will then attempt to register again, causing congestion at the S-CSCF.

This is one of the key reasons that registration failures should not result in a registered device from becoming de-registered. If a device fails registration, an error response should be returned to the device, but the existing registration remains intact until it reaches its end of life, or the device sends a legitimate re-registration. This will prevent network congestion from attacks using the registration process.

Integrity

Integrity ensures that what was sent is the same as what is received. This is considered an aspect of security because a man-in-the-middle attack, for instance, could intercept a SIP message and alter its contents (redirecting responses back to the perpetrator, for example).

Message body tampering is an example of where integrity is needed. By preventing this from occurring, an operator can be assured that masquerading will be made more difficult. To protect the integrity of a message, communication with the user device is conducted through secure ports. These ports are assigned after a user device successfully completes registration.

The P-CSCF assigns a secure port (other than 5060/5061) and uses this port for all communications with the device until it registers a new location. By using the secure port, all entities within the home network can be assured that the message is being exchanged between two known and trusted entities, and that the device has already been authenticated.

If a message is received by the P-CSCF that is not a *REGISTER* message, and it is received on port 5060/5061, the P-CSCF discards the message. Today's VoIP networks do not use security association in most cases, allowing hackers to gain access into the various entities within the network and open these ports to accept all SIP traffic. Security association eliminates this vulnerability.

This also prevents replay of a message. Replay is where a message is captured by the perpetrator by eavesdropping and recreated for the perpetrator's own use. Usually some parameters are altered, such as the destination address, so that the perpetrator can gain access to operator services masquerading as a legitimate subscriber.

For example, a hacker could capture SIP *INVITE* or *REGISTER* messages and then use these messages to obtain services for his or her own use. The hacker could change the location address in the *REGISTER* message and change the destination address in the *INVITE* message, immediately gaining access to services it otherwise would not be allowed to use.

Privacy

Privacy allows subscribers to remain anonymous to various networks. While within the home network, the network always knows the public and private identities of a subscriber (otherwise services are not provided). But transiting networks are not typically provided this information if privacy has been invoked.

When privacy is invoked, there is a means for the operator and any partner operator to know the origin (defined by the private user identity) of a message. This is another step toward preventing users from hiding their identities from the service provider. Of course if a subscriber does succeed in hiding behind an anonymous server, for example, the service can be denied by the operator (based on the inability to identify the private user identity).

The purpose of privacy is to protect the identity of subscribers from transiting networks. This would prevent capturing identity data while transiting various networks to reach the destination network. Of course the destination network will need to establish the identity of the subscriber prior to providing any services. The destination (or terminating network) would still have the ability to ascertain the identity of the subscriber, establish that the subscriber has been authenticated, and determine the services that the subscriber is authorized to use.

Where privacy presents a challenge, this is lawful intercept. If a law enforcement agency serves a subpoena on a transiting carrier, for example, the transiting carrier may not be able to provide the true identity of a subscriber, simply because they will not typically have access to this information. Only the home or the terminating network will have this visibility. Of course, the transiting carrier can simply identify the home network, and law enforcement should then be able to obtain the data they need from the home network.

Non-Repudiation

Non-repudiation prevents subscribers from denying they participated in a session, when they actually did. This is much simpler to accomplish when an operator is using monitoring systems to capture all of the SIP messaging. By doing so, the operator is able to identify each subscription that accessed a service and, given the extensions added to the SIP protocol for authentication and authorization, verify that subscribers who accessed those services are who they say they are.

This is one of the reasons many operators have begun using signaling as an audit source. Previously, operators relied on call records generated by each of the network elements for billing and billing verification (auditing). However, when they went into a billing dispute with another operator, they found it difficult to prove their cases with call detail records generated by the network entities.

Over the last five years, many operators have begun using the SS7 signaling data to generate call detail records for the purpose of auditing their billing functions. This has become a popular practice for a number of reasons. The primary reason is that signaling is the best reference data for auditing billable transactions. Signaling is what sets up sessions and tears them down. This makes signaling indisputable when it comes to arguing if a service was delivered or not.

In order for a call (or a session within the IMS domain) to be connected, signaling data must be exchanged between the originating party and the terminating party. In the legacy public switched telephone network (PSTN), signaling is used to connect each and every circuit needed to establish an end-to-end connection.

In the SIP domain, signaling is used to negotiate the bandwidth and resource requirements for a session. Likewise, SIP is used to reserve a port at each of the devices and reserve that port for the purposes of maintaining a dialog with another entity. It's the same concept, just a different form of signaling. What operators discovered was that when they used signaling for their billing disputes, simply because of the nature of what signaling does, the data was indisputable. Without signaling there can be no connection or session.

This also works to prove subscriber usage. When a subscriber denies connecting or downloading content, the signaling can be used to provide an audit of the transaction and an indisputable record of the entire transaction. Any operator who is not using signaling as a part of its auditing of service delivery is most likely losing money.

Here is how it works. If a subscriber wants to download a music file, SIP messages (an *INVITE* initially) must be exchanged with the music download server. Once the server has authenticated and authorized the subscriber for the download, the server can then send the file using the File Transfer Protocol (FTP). At the conclusion of the download, SIP is used to release the session.

You can read more details about establishing a session in previous chapters. The main point here is that SIP identifies each and every transaction in the network, and it confirms whether or not service delivery was successful. If there was a problem with service delivery, SIP will return an error code indicating that failure.

This is why SIP (or any other form of signaling) is so important to operators when auditing subscriber billing and preventing repudiation. It's surprising that more operators have not implemented this simple process already.

Access Security

The first step to securing any network is ensuring only those subscribers authorized to use the network services are provided access. This requires a few procedures and certainly can be challenging for any operator.

The most important is authentication. This process is required to ensure the subscription is valid (they are who they say they are), and the device requesting services is authorized to access the network. How this is accomplished is explained in the next section.

You must understand, however, that the network can only authenticate and authorize the device attempting access. There is no means of authenticating the person using the device, unless the device itself employs some form of biometrics. The network assumes that the use of the device is legitimate and allows access to the device (it is assumed that the device has been adequately protected).

The ultimate security, of course, begins at the device level, ensuring that the device itself is protected from unauthorized usage. This is difficult to enforce because most people do not want to be bothered with things like biometrics and passwords. If there were a way to authenticate the user of the device at the device, without too much inconvenience to the subscriber, security would be a lot easier.

When access to a device is compromised, the attacker has full access to the network. However, there are other methods to gain access into the network without using the actual device itself. Masquerading as a legitimate subscriber allows hackers to use their own devices to gain access into the network.

There are methods that can be used to prevent man-in-the-middle, spoofing, and other forms of attacks where hackers gain access through other network entities and masquerade as an authorized subscriber. These are a little more difficult to prevent, but again not impossible.

Some of those processes require strict routing in the network (rather than loose routing as often implemented in VoIP), as well as encryption. Most operators shy away from these methods because of the complexity it adds to the network. However, when one considers the added security, one must weigh the cost of losing revenues and degradation of services vs. network complexity.

Wireless certainly brings about many of these concerns, as well as a few that wireline operators do not have to worry about. There is a lot of concern over the air interface within a wireless network, especially with WiFi and WiMAX technologies. Devices exist today that make eavesdropping over the air interface much simpler, and therefore wireless operators are looking into various ways to encrypt and secure the air interface to prevent unauthorized eavesdropping.

The GSM and CDMA operators have already implemented encryption over the air interface for exactly this reason. This has resulted in a sharp decline in handset cloning, where hackers would "eavesdrop" on the air interface trying to capture the IMEI and IMSI of a device, and then copying that data for use in their own devices. Encryption can have the same results internal to the network as well, if implemented properly.

Encryption prevents unauthorized users from gaining access into the network. There should be a mechanism for authenticating everyone who enters the network. Knowing that subscribers really are who they say they are requires authentication. However, authentication each and every time a subscriber accesses the network is not really efficient. It would require a lot of network resources and tie up the S-CSCF. Once a device is registered in the network, it has already been authenticated. There should be a means of identifying that the subscriber has been authenticated, so that the process is not repeated every time that subscriber establishes a session.

SIP provides an additional header that is inserted into the SIP message after authentication. The *P-ASSERTED-IDENTITY* is a trusted header indicating that the authentication process has already been completed, and the URI contained in this header has been verified as the true identity of the subscriber. This header is used by all IMS entities within a trusted domain instead of the *FROM* header, which is not to be trusted.

The 3GPP added this header because they recognized it was not efficient to authenticate for every session. It can also be used to identify the true user identity of a device where the *FROM* header indicates something else.

If a subscriber has implemented privacy features (as indicated in the SIP message), the *P-ASSERTED-ID* is still provided within the host network, but it is removed prior to sending to other networks. This means that a SIP message coming from another network may not contain a *P-ASSERTED-ID*, presenting an authentication challenge to the operator. In this case, authentication is not necessarily needed to determine if the subscriber should have access to the network, as this is taken care of through other procedures. Rather, the concern is whether or not subscribers are who they say they are, and identifying a subscriber to the destination party.

When a message is received with a *PRIVACY* header with the token "*ID*," the entity removes the *P-ASSERTED-ID* header from the message and forwards the message to the destination. If the token is set to "*NONE*," then no privacy has been requested. It is then up to the operator policy to determine if the header is to be provided outside of the network.

When an IMS network receives a SIP message from another network, and the *P-ASSERTED-ID* is missing, the I-CSCF should then authenticate subscribers to verify they are who they say they are, and they indeed have authorization to access the network.

The HSS serving a subscriber verifies if the subscriber should have access to the visited network, and also provides the authentication needed.

The P-CSCF also plays an important role in safeguarding the network against unauthorized access. For this reason, all subscriber devices are forced to first connect with the local P-CSCF. No entity aside from the home P-CSCF and I-CSCF functions should be allowed direct access to the S-CSCF in the network, to prevent against masquerading and other forms of attacks. The P-CSCF protects the S-CSCF and the HSS.

One of the P-CSCF functions is verifying the integrity key *(IK)* each time a registered device accesses the network. If the *IK* is invalid, the P-CSCF discards the message, preventing the device from getting any further.

The device and the P-CSCF maintain two security associations, one for incoming and one for outgoing ports. These security associations are established after registration and are reserved for all communications between the registered device and the concerned P-CSCF. By maintaining a security association, the P-CSCF can prevent attacks from other devices masquerading as legitimate subscribers.

An association begins once registration has been completed using the SIP port 5060 (or 5061). These are the commonly used ports called out in operating systems for all SIP sessions. However, in the IMS, these ports are only used for registration. Once registration is completed, the device is assigned a different port known only to the device itself and the P-CSCF.

IPsec manages the associations on these ports. When the security association is established at registration time, the IP address at the IPsec layer and the IP address at the SIP level are compared to ensure they are the same. The IP address at the SIP level is found in the *VIA* header, but if the *VIA* header contains a SIP URI, it must be resolved to its IP address prior to continuing on. This requires a query to the Domain Name Server (DNS) by the P-CSCF.

The ports that are assigned as part of the security association are dedicated to the subscriber device until it changes its registration. These ports cannot be assigned until after the subscriber device has been authenticated and registration is complete (see Figure 6.1). Only six security associations per device can exist at any one time.

This means that the standard SIP ports 5060 and 5061 are not used for authorized secure communications. These ports are only used for registration and error messages. If an INVITE is received on port 5060 or 5061, the P-CSCF will discard the message as unauthorized. This simple security procedure would have prevented many VoIP security breaches if it had been implemented in VoIP networks today.

To prevent man-in-the-middle attacks, the P-CSCF will check the route taken by the message as recorded in the *RECORD-ROUTE* headers. When a subscriber registers with the network, the *RECORD-ROUTE* headers are used to create a route list for the subscriber. This route list is stored in the HSS as part of the subscription's registration and is sent to the S-CSCF as well.

The P-CSCF maintains a record of the route when it receives the 200 OK response to a *REGISTER,* and it uses this to verify that any request from the subscriber follows the same route. This ensures that a message is not copied or rerouted by hackers and then used to create duplicate messages (cloned) for accessing the network (replay attack).

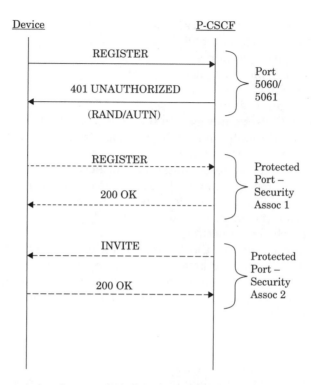

Figure 6.1   Security associations between the subscriber device and the P-CSCF

The route list is maintained throughout the period of registration. If the subscriber moves (or changes the IP address), then the subscriber device must be registered again. The new route list is created during this new registration process.

If a hacker had copied a SIP request and used it to create a "cloned" message for unauthorized access into the network, the route used would be different than the real subscriber (most likely). This then makes it simple to verify subscribers by location alone.

If the P-CSCF receives a request with a route other than the recorded route, it will change the routing headers in the message and forward the SIP message through the authorized and registered route list. This would then result in the request/response being sent to the authorized subscriber and the appropriate rejection being returned.

This also prevents session hijacking where an *INVITE* is intercepted by a hacker, who then sends a 3*xx* response. This form of attack redirects a request to the hacker's server, which then assumes control of the session. This can also be used in multimedia sessions where voice and video are used, for example. In these cases the hacker could redirect a portion of the call, since each portion of the call (the video being one portion) requires its own session.

Transiting networks obviously are not going to know the route taken in the originating network, and therefore will have to establish their own route lists. When a message is sent to a transiting network, it should also be authenticated and a route list established at that time. In other words, each network is responsible for its own authentication and security procedures. No network should be considered as 100 percent trusted. Certainly today there are many operators responsible for fraudulent and unethical practices, sending traffic into legitimate operators' networks.

Another means of securing access into the IMS domain is by using the Interrogating-CSCF (I-CSCF) as a gateway into the network. All other networks gain access to the home network through the I-CSCF function. The I-CSCF then blocks network particulars from other interconnecting operators.

For example, the addresses of the various network entities can be hidden from other networks using topology hiding. This function encrypts headers such as *VIA, RECORD-ROUTE, ROUTE,* and user identities. Only networks with business agreements in place are given the encryption keys so that they can decrypt these headers and interoperate with the home network.

By using this function, the home network prevents other networks from determining the network topology, as well as the number of hops necessary to reach network resources. This information could be used to launch denial-of-service attacks and even unauthorized access to network services. The I-CSCF provides protection for the HSS from unauthorized networks.

Encryption

Encryption is an important function within IMS networks. Without encryption, SIP headers are readable (in plain text) and can therefore be captured by network sniffers and other devices. This enables hackers to use the information contained in these headers to "clone" messages for gaining unauthorized access to the network.

I have already given many examples of where encryption can be used to prevent hackers from stealing identities, and to prevent rogue operators from learning the network topology of a competitor in an attempt to gain unauthorized access to network resources (such as the HSS). Encryption is important internally to the network as well as externally. Hackers have the ability to "listen" to traffic within a network and capture data about subscribers and sessions they can then use for their own access.

One hacking method is to hijack registrations. This is done by copying headers from legitimate SIP messages, copying the public and private identities, and registering on behalf of the real subscriber. All messages are then redirected to the hacker.

Another example of why encryption is necessary is to prevent session hijacking. This is accomplished when a hacker intercepts an *INVITE* and sends a 3*xx* response redirecting the request to their server. Session hijacking can be prevented partially through encryption (the contents are no longer readable by hackers) or through route verification at the P-CSCF.

There are downfalls to encryption as well. When the SIP message is encrypted, only those devices with the cipher key will be able to decipher the message. These are typically the proxies in the network (call session control functions within the IMS). Monitoring systems used for performance management may not have this capability.

Likewise, lawful intercept may also become a challenge. Without encryption, however, there are many more consequences.

Transport Layer Security (TLS) and IPsec are two methods recommended by the 3GPP for encryption. IPsec works within a trusted domain, providing encryption between trusted entities. This prevents eavesdropping by entities between call session controllers, for example. IPsec is implemented within the operating system of the various network nodes themselves working at the lower layers.

This includes between CSCFs as well. IPsec should be used between all entities within an operator's network. However, IPsec is not useful when connecting to another network. This requires TLS.

TLS works at the TCP (transport) level. It works best when connecting to two unknown entities. For example, when transporting a message through multiple service provider networks, TLS is used at the transit level. IPsec is used within the trusted network entity.

TLS does not work well end-to-end. It is only useful when used in transit networks. TLS can be used to prevent hackers from intercepting *REGISTER* messages and obtaining subscribers' credentials in these transit networks, though. This is the strength of TLS.

TLS provides authentication, integrity, and confidentiality as well and is recommended by the 3GPP standards as a means of encryption for IMS in transit networks. Of course it presents some challenges for the transit operators, because they will not be able to see some of the encrypted headers, so they must enforce their connection agreements to ensure they are providing access to their networks to authorized and trusted operators.

Authentication and Key Agreement

I have been talking a lot about authentication of a subscriber, but I haven't really explained how a subscriber device is authenticated. For a network to be able to authenticate subscribers, there needs to be some exchange of security information between the subscribers' devices and the network. This is accomplished through the use of security "keys" that are programmed into the subscriber devices and shared with the network when a subscriber activates their device.

This is analogous with how GSM networks exchange authentication data between handsets and the HLR today. However, this mechanism also works within a wireline model as well, as long as the device has some fashion of ISIM application that is inserted into the device providing the authentication and subscription identifiers. IMS defines the use of UICC within each device hosting the ISIM application.

When a subscriber purchases a device from his or her service provider, the device is already equipped with a UICC hosting the ISIM application. The operator has therefore already provisioned the ISIM with the appropriate authentication data.

If the subscriber purchases a device from someplace else, that subscriber still has to have the proper UICC and ISIM application from his or her home service provider. This is the only way for authentication to work properly.

The service provider's Home Subscriber Server (HSS) is where security keys for each subscriber are kept. When a subscriber activates his or her device and the registration process begins, the S-CSCF assigned to the subscriber queries the HSS for the security data. The S-CSCF will then "challenge" the subscriber device for this information.

The challenge is talked about several times in this book and is one of the fundamental differences between traditional (if there exists such a thing) Voice over IP (VoIP) networks and the IMS. Many VoIP networks today do not challenge subscriber devices when they are accessing the network. Indeed many of the VoIP fraud cases this author has researched were a direct result of no authentication within the network. Authentication is paramount to ensuring services are accessed only by those authorized to use the network.

There are a number of ways that this can work. We are talking about one example where precalculated security keys are stored in the device and the HSS in the network, but there can be automated methods where algorithms are implemented to dynamically assign security keys as devices register. This would work much like security tokens used for accessing data networks today. Each private user identity is associated with a security key and at least one of its public user identities as well. In reality, all user identities should be associated with security keys. They are stored in the HSS, and in the ISIM of the subscriber device.

When the subscriber sends the *REGISTER* message to begin registration procedures, the I-CSCF assigns an S-CSCF. The S-CSCF acts as the registrar in the SIP domain and will challenge the subscriber device to authenticate the subscriber. The S-CSCF will check the *AUTHORIZATION* header to see if this subscriber has been through the registration process already. If the header contains the *INTEGRITY* parameter with a value of *NO,* then the S-CSCF will challenge the subscriber.

This is not a fail-safe method, so it is usually advisable for the network to proceed with challenging the subscriber device even if the *INTEGRITY* parameter does indicate a previous registration. Any *REGISTER* coming from the device should be treated as suspect, to prevent unauthorized access to the network.

The S-CSCF responds to the first *REGISTER* message with the response 401 Unauthorized. Before sending the response, the S-CSCF then queries the HSS for security credentials. The HSS then sends via DIAMETER the random number *(RAND)* and authentication token *(AUTN),* as well as the expected response *(XRES).* The *RAND* parameter contains the cipher key *(CK)* and the integrity key *(IK).* The S-CSCF then sends this information in the 401 Unauthorized challenge to the device. The device then compares the *MAC* in the *AUTN* header with a value stored within the device (in the ISIM).

When the device receives the 401 Unauthorized response, it uses the *MAC* parameter in the *AUTN* header, calculates *XMAC,* and verifies the two matches. The value is then sent along with other credentials in a new *REGISTER* message carrying the same *CALL-ID* as the first *REGISTER.* The *CALL-ID* is the key that lets the S-CSCF know that this is related to a previous challenge.

The *SECURITY-SERVER* header is also sent in the challenge. If this is missing, the device abandons registration and starts all over again by sending a new *REGISTER* with a new *CALL-ID.* The *SECURITY-SERVER* header identifies the security methods supported. It also identifies the IPsec algorithm to be used, and security association parameters.

A security association is established between the subscriber device and the P-CSCF as mentioned earlier. Remember our discussion earlier regarding secure ports at the P-CSCF. These ports are part of the security association. The P-CSCF uses the data provided in the *REGISTER* message to establish the security association after registration has been successfully completed. The security association helps prevent against message tampering and replay attacks.

The S-CSCF will be expecting a new *REGISTER* message containing the necessary credentials. When the device receives the `401 Unauthorized` message, it creates a new *REGISTER* message containing the same *CALL-ID* as the previous *REGISTER*. This is so the S-CSCF knows that this is in response to the S-CSCF challenge.

The S-CSCF will then query the HSS for authentication keys for the subscription. The HSS will then send to the S-CSCF a random number *(RAND)*, expected response *(XRES)*, cipher key *(CK)*, integrity key *(IK)*, and authentication token *(AUTN)*. The device will have the proper criteria as well, stored in its ISIM (IM Services Identity Module).

In the new *REGISTER* message the device adds the SIP header *AUTHENTICATION* with the parameters *RAND, AUTN, IK,* and *CK*. These parameters are calculated by the device using data provided by the S-CSCF in the `401 Unauthorized` response. Only the operator and the device know the algorithm used to create the correct response based on the given data, so if another device from another operator attempts to use the data provided, the calculation will be incorrect, and the authentication will fail.

If the subscriber is roaming in another provider's network, the visited network S-CSCF will send the challenge and query the HSS of the home network to determine if the subscriber is legitimate and has permissions to allow access to the visited network. The S-CSCF in the visited network becomes the registrar for the subscriber while that subscriber is roaming in the other network.

As mentioned before, authentication is absolutely paramount to any security initiative and should be the first step to any security plan. The IMS provides the mechanism for authentication within the IMS domain, which should solve some but not all security problems. There are still other ways in which unauthorized access can be gained to an IMS network.

The biggest challenge will be for wireline operators to provide some fashion of ISIM that can be used to enable any device provided by the subscriber, as it is most likely that subscribers will purchase their own devices from multiple sources for network access. These may be SIP phones, PDAs with network access cards, or similar types of data devices.

Network Domain Security

To understand security measures for the IMS, you must first understand the concept of a "trusted domain." This is quite simply a network domain where the "owners" trust one another, or there is only one "owner." For example, in telecommunications networks, your service provider owns their own network. This network will interconnect with many other networks to offer global service. However, those other networks belong to other companies, or "owners."

Through special agreements and arrangements, the various network operators build a trust among each other. This trust allows for the exchange of certain data and information about the subscribers, routing information, and addresses for resources within each network. The IMS is no different and uses the same "trusted domain" model.

Securing the network boundaries through the use of gateways is a critical part to any network security and is no different in the IMS. The P-CSCF acts as the first point of contact for any user of the network and provides some important security functions. Forcing routing based on route lists established during registration (strict routing) is one form of security that the P-CSCF can provide to the network.

The I-CSCF acts as the gateway between two IMS networks. This means that the I-CSCF has the responsibility for determining what information is shared between two networks. The I-CSCF may support topology hiding, for example, and may implement encryption to prevent unauthorized access to proprietary information. The I-CSCF should be implemented with all possible security options, because this entity is what provides access to your network and its resources.

The S-CSCF is the registrar for the network and maintains all registration information for each assigned subscriber. As discussed in other chapters, the S-CSCF can be assigned to a subscriber in numerous different ways, but for security purposes it may be more desirable to assign a subscriber to a fixed S-CSCF rather than dynamically assign the subscriber each time that subscriber registers.

The HSS should also be heavily protected, since all subscriber data is stored within the HSS. The S-CSCF provides access to the HSS and should be used to prevent unauthorized access through authentication.

We have already talked about the various roles of the CSCF within the network and how it is used for security. The P-CSCF protects the S-CSCF and the HSS, while the I-CSCF protects the network from other networks.

Routing within the IMS should be more controlled than what is found in other networks. For example, in VoIP networks, loose routing is typically used. Loose routing allows the various proxies to determine the best route for requests and responses to take, based on current traffic demands and various other conditions.

In the IMS, loose routing represents security issues, which is why strict routing is recommended for routing between the CSCFs in the network. One of the first steps in IMS is to enforce the use of *ROUTE* and *RECORD-ROUTE* headers. In SIP domains, this is an elective, but in IMS this is required. Using the *RECORD-ROUTE* header to record the path taken during registration, and then storing this information as part of the registration, ensures that messages are not inserted from other parts of the network to hijack a session, or change a registration.

The *RECORD-ROUTE* and *ROUTE* headers ensure that all requests/responses are forced through the same proxies (call session controllers in the IMS) acting as gateways. How these are handled between networks should also be taken into consideration.

We talked about this process with the P-CSCF where this is enforced. Since the P-CSCF is the first point of contact within the IMS, it only makes sense that the P-CSCF be mentioned here as a means of protecting the network domain.

The information that gets exchanged between networks will depend on whether or not the other network is a trusted network and has formal agreements in place. If no agreement exists and the other network is not a trusted domain, then measures must be taken to ensure the IMS network is secure. This means that procedures must be put into place to guard certain subscriber information elements from being exchanged with non-trusted domains.

Topology hiding can be used to prevent network information that could be used to understand the topology of the network from leaking to other networks. The I-CSCF acts as the gateway into the network and strips any headers from outgoing SIP messages that would contain addresses of internal IMS entities. These addresses could be used to determine the number of S-CSCFs in the network domain, for example.

Usually this information would be found in the *ROUTE* and *RECORD-ROUTE* headers. Collecting this information could disclose to an outside network the number of nodes within the network domain route, network capacity (if used in conjunction with "test" traffic), and even S-CSCF capabilities.

Topology hiding is one means of preventing call session controller functions from being compromised. By hiding the addresses of these entities within a network, it prevents unauthorized personnel from learning the addresses and attempting to "hack" into the systems. Once compromised, incoming *REGISTER* messages could be redirected to a "phony" registrar in another network.

The rogue server could also send a response back (`301 Moved Permanently`) that would reroute all subsequent requests/responses to the rogue network. This is being played out today with Internet services, with rogue individuals hijacking HTTP servers in the network and rerouting all messages from these servers to their own servers.

However, this information is often needed in order to terminate calls within another remote trusted domain. To allow for this information to be read by these networks, encryption is used on the headers providing proprietary data. Only trusted networks are able to decrypt these messages using the encryption keys from the originating network. The I-CSCF performs the encryption prior to forwarding messages outside of its own domain.

When encryption is used, the *VIA, ROUTE,* and *RECORD-ROUTE* headers are all encrypted. This prevents other networks from learning the addresses within the operator's network and learning the topology of that network, while also preventing eavesdropping. The challenge it presents, however, is in the area of performance management.

Performance management solutions typically rely on some fashion of probes or sniffers in the network to copy and collect signaling messages in the network, and use this information for calculating statistics and generating reports on the overall health and performance of the network. If the SIP messages have been encrypted, this presents a challenge to this method of collecting signaling messages and will then require an alternative solution. Lawful intercept is challenged here as well, as many times these performance management solutions are also used to support lawful intercept.

When topology hiding is being used, the P-CSCF does not know how to decrypt the messages (it does not have the cipher key). Only the I-CSCF has the cipher keys, and therefore the receiving network must also have an I-CSCF providing topology hiding in order to decrypt these messages as they are received. The receiving network is considered in this case the terminating network. The terminating network must be able to read the headers for routing responses back to the requestor in the originating network. Transiting networks do not need to see these headers but rely solely on the *request-URI* for routing of the message. Remember the *request-URI* is located in the first header of the message, where the method is identified, and is used to route the message to its final destination (or the next hop in the network).

Topology hiding is also used to screen SIP messages and strip other information that is of local significance only, such as billing information and IP addresses. Hiding this information is useful for a number of reasons. Charging information is never shared between two networks, and therefore the I-CSCF must ensure that the headers providing information regarding how the network was accessed *(P-ACCESS-NETWORK-ID)* and all charging headers are removed prior to forwarding the SIP message to another network.

In any network, it is always a good idea to maintain border security and not disclose a lot of details about the nodes within the network. In fact, routing of all traffic through gateways allows operators to publish a finite number of addresses (or in the case of IMS, an Anycast address). The gateway or topology hiding function then knows the internal addresses and is responsible for routing messages from other networks to the appropriate entities within the network domain.

Online and Offline Charging in the IMS

The most important function within the IMS is charging. It is by charging that operators are able to collect revenues for the services they offer, and what allows them to continue to fund the development of their network and future services. While the model in place today for how operators charge for services (a minutes-of-use model) is rapidly changing (to a content/services rendered model), there lacks a means today for supporting the charging of many different types of services within one system or platform.

The IMS provides a fresh opportunity for operators and vendors alike to implement a major change within their charging architectures. Literally everything that a subscriber does is tracked and recorded within the IMS, which is one of its principle advantages from a revenue perspective. If a subscriber places a voice call, every aspect of that call is captured and recorded, much as it is today with one significant difference.

Every other transaction by that subscriber is captured as well, such as e-mail, instant messaging, exchanging files with other subscribers, and even initiating video conferences. Why is this different from today's model? Because today this would have to be accomplished by several different platforms capturing data from many different entities within the network. Worse, many different billing formats are required to support multimedia. There lacks within the industry one format for billing records to fit any type of communications service.

This is perhaps the single most daunting issue operators face when moving to a new architecture such as IMS. How do you convert all of the back-office systems and billing architectures in such a way as to support the many different services you are implementing, without deploying multiple systems and billing formats? The IMS provides a single reference point where all of this data is captured, correlated, and distributed to billing systems for charging. But more importantly, the IMS introduces a new billing format supporting all forms of media, enabling operators to implement one billing platform for all of their media services.

Introduction to Charging

Charging in an IMS is supported through the DIAMETER protocol and a whole new charging architecture. The various network elements within the IMS that are responsible for charging, support the DIAMETER protocol through various interfaces so that charging detail records (CDRs) can be collected for any services rendered. The DIAMETER protocol is used by all entities to communicate billing details (but not the CDRs themselves). The DIAMETER protocol then communicates usage data for the various services provided by the reporting entity.

This is a departure from earlier models where the switches supporting voice services also created a *call detail record*. The call detail record was formatted using one of many standard variations, and forwarded to a mediation platform that "normalized" all of the disparate billing formats into one consistent format supported by the billing presentment system.

This new charging architecture consists of functions that are dedicated to charging. These functions may be integrated into other network elements, or they may be stand-alone devices. Rather than create the CDR at the network entity itself, and then hope that the CDR format is supported by the billing system, the network entity coveys billing details to another centralized function responsible for creating the actual charging record for the billing presentment system.

The processes and procedures for these functions are still being defined, but there is enough defined in the 3GPP standards to understand how charging will work, and how these various functions interact with other IMS entities.

Charging Architecture

There are two aspects to charging in the IMS; online and offline charging. Offline charging is the charging for services after the service is delivered. This is also known as post-paid charging. Online charging is the charging for services prior to service delivery. This is known as pre-paid charging. The mechanisms used to support both online and offline are different.

For online charging, there is an online charging system (OCS), while for offline charging there is a charging data function (CDF) and a charging gateway function (CGF). Both online and offline charging rely on a charging trigger function (CTF) residing in each of the network elements to capture charging data.

The CTF is provisioned with the addresses of all of the CDFs and OCFs within its own domain. It is to these addresses that the trigger function will forward charging data using the DIAMETER protocol. Think of the trigger function as the software within the network entity that provides the actual usage data.

For example, if an operator provides a video service, a voice service, and a messaging service, there will be three distinctly different billing formats provided. For the voice service, the media gateways will need to provide the minutes of use (if this is still supported) as well as the number called. This information will be necessary for rating of the call.

For the video service, the video server will need to provide the billing system with the video downloaded (or played using a streaming service) and how long the video was watched. This also is sent using the DIAMETER protocol to the function responsible for creating the actual CDR for rating and billing presentment.

The SMSc (responsible for the delivery of Short Messaging Service) sends details about the number of messages received or sent by any subscriber, again using the DIAMETER protocol. Without this mechanism, each entity would be responsible for sending a CDR containing very different information that would then have to be defined and standardized based on each and every type of service.

What the IMS charging architecture brings to the table is a means of using the DIAMETER protocol to communicate this data between all of the various network entities without having to normalize the actual record itself at that entity. The CDR then gets built within the core of the network, using data sent by each of the entities. Any changes in rating or changes made to the service plans can be made easily without having to make any changes at each of the network entities.

Charging information is not sent outside of a network domain, unless the operators have a previous agreement to act as "trusted" domains and share subscriber and billing data. However, there are provisions within the standards that do allow networks to share certain billing data between each network.

The addresses listed in the CTF may be weighted or prioritized to allow for distributed routing. This allows multiple CDFs and OCFs to be provisioned in the same domain for load sharing and failover. If the first CDF or OCF listed in the routing list is busy or unavailable, then the CTF has the option of sending to the next address in the list. This is another advantage to the architecture: allowing operators to build their charging networks where backhauling all of the billing data does not have a significant impact on the network backbone.

There are a number of interfaces used between the various elements within the IMS charging architecture. Each of these interfaces supports specific functions within the charging process and is shown in Figure 7.1.

The network elements know to which interface to send charging information from information received in the SIP message. The SIP header *P-CHARGING-FUNCTION-ADDRESSES* provides the address of the CDF in the *CCF* parameter, and the address of the OCS in the *ECF* parameter. This data is then used by the receiving entity for routing of the DIAMETER message containing the actual billing data.

Operators may choose to provision their application servers with these addresses rather than use SIP for the addressing. This means that the application servers would then rely on provisioned routing tables rather than the SIP headers to determine routing for their DIAMETER messaging. This is allowed for, and for some applications it may be a better approach. However, this is implementation specific and the network supports either implementation.

Online Charging Architecture   The online charging function within the IMS depends on the charging trigger function (CTF) resident in the various elements to provide charging data to the OCS. Remember that on the front end of the network entity, SIP messages

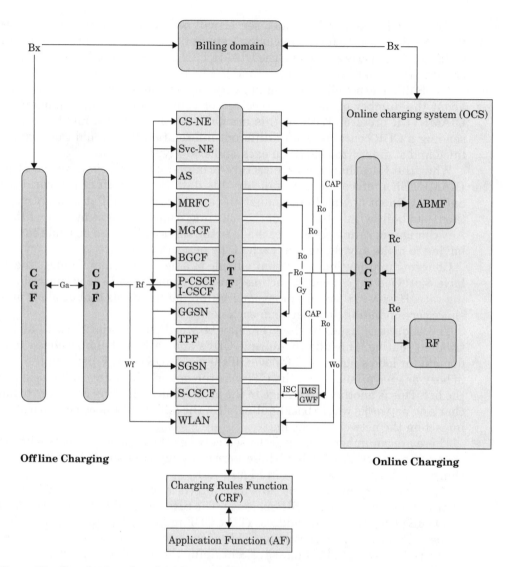

Figure 7.1 Charging interfaces for online and offline charging

are being provided with regard to a specific service request. Using our previous example of a voice call, a media gateway controller (MGC) would send a SIP *INVITE* to the local P-CSCF for routing through the IMS. The P-CSCF would then take specific headers from the SIP *INVITE* and use this data to create charging data. This data is then sent using the DIAMETER protocol to the specific address in the SIP *P-CHARGING-ADDRESS* headers for the OCS.

The OCS has an OCF used for collecting the billing data. This interface then receives the DIAMETER message and uses this information to rate the service being provided.

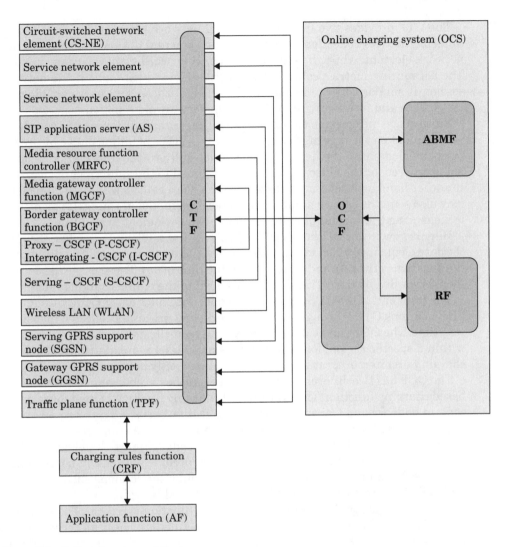

Figure 7.2 Online charging architecture

The OCF interfaces to the rating function (RF), where the session is rated and the account balance management function (ABMF), where the subscriber account is verified for adequate funds (see Figure 7.2).

Since online charging takes place in real time (services are paid for prior to service delivery), there is no need for CDRs. The account balance is charged prior to the service being delivered, which means that the ABMF function must not only verify the account balance but make sure that the account is decremented for the service about to be delivered.

The CTF provides two functions within a network element. The first function is accounting metrics collection (or usage measurements) of services used at that specific network element, while the second function is forwarding of the usage measurements. The accounting metrics collection can be distributed throughout the network, so if a session is supported by multiple devices, each of those devices is capable of collecting usage data and forwarding that data to the forwarding function.

Since usage measurements (or accounting metrics, whichever term you choose to use) is dependent on the type of network element, the type of billing data captured will differ from one element to the next. A video service will provide information about the video being ordered, or the video that is being streamed to the subscriber device. An MGC, on the other hand, will send minutes of use for a voice call, as well as the dialed number. It may also send the calling number for proper jurisdictional rating. Its principle function is measuring the consumption of services at that element in real time.

Here is how it works. When a call is generated at a telephone switch, the CTF collects the data elements needed to determine usage. For a call, this would be the number dialed, the duration of the call, and possibly even the time of day. These usage measurements are then forwarded by the CTF using DIAMETER to the online charging system (OCS).

The front end of the OCS is the online collection function, or OCF. The OCF interfaces to the various CTFs throughout the network, and processes collected measurements in real-time. The OCF can be distributed throughout the network, interfacing to the CTFs within a specific region, for example. This provides a more distributed approach to the charging architecture, rather than a centralized-only approach.

The OCF has two different functions: an event-based charging function and a session-based charging function. The event-based charging function is used for services such as content delivery and transactions with application servers. The session-based charging function is used for services such as voice calls and data transfers.

It is probably safe to describe event-based charging as the charging for sessions where duration or bandwidth cannot be calculated, while session-based charging does calculate duration and bandwidth consumption. User sessions such as voice calls and GPRS PDP contexts are examples of session-based charging.

Since this is prepaid charging, the real-time function is critical. The OCF must determine if there is enough money in the subscriber account prior to service delivery. This means that the CTF must be able to cancel any service delivery request if the account balance is depleted, or terminate service delivery in the event the account balance is depleted while the session is in progress.

The service must first be rated through the rating function (RF) so that the correct charges can be calculated, and then the account balance must be verified through the account balance management function (ABMF). All of this is controlled and managed through the OCF prior to service delivery.

The rating function will rate according to the volume of data being sent (bandwidth consumption), the duration of the session (length of the call), or content delivery (which includes any downloads or messaging).

Once the event has been rated, the OCF then sends the charging event to the ABMF to determine the account balance. It should be noted here that there are no CDRs generated for these charging events, because the service delivery is dependent on the current account balance. The service is paid for before the service is actually delivered. However, there are provisions for the creation of CDRs using the same components used in offline charging for CDR generation, namely the CDF.

This could be useful for auditing of services, as well as monitoring and marketing intelligence. These applications rely on some form of CDR for creating reports and trending of service delivery.

The OCS does not represent the prepaid system itself. The OCS interfaces with the prepaid platform via the ABMF. This allows operators to maintain existing prepaid platforms and adapt them for use with IMS networks as they transition their networks. There may need to be a box supporting the Bx interface in order for the prepaid platform to communicate with the OCS in the IMS domain.

The interface to the prepaid platform is actually a charging detail record (CDR), enabling existing systems to easily evolve by supporting the new CDR format. These formats are still being defined today.

Offline Charging Architecture   Offline charging operates differently than online charging. First of all, offline charging requires the generation of CDRs in order for billing to be applied at the end of a billing period. This is not the case with online charging because online charging happens in real time.

As with online charging, the charging trigger function (CTF) within each network element is the critical link in the process (see Figure 7.3). The CTF must collect all usage data within the element, correlate the data and send it to the forwarding function for processing using the DIAMETER protocol.

The CTF then interfaces to a charging data function (CDF). The CDF is responsible for collecting charging data from all of the CTFs much as the OCF did in the online charging architecture. The principle difference is that the CDF creates CDRs.

The CDF collects charging events from each of the CTFs in the network and uses those events to create a single CDR or multiple CDRs. In other words, the CDF will create a CDR for each individual event, or it will create a single CDR for multiple charging events, provided all of those events occurred at the same network element. The events can be from different types of sessions, for example, a voice session combined with a video session (a video conference call) would be two different charging events but could be recorded in one CDR. It would be up to the operator to configure the CDF rules.

The CDF is most likely a stand-alone function, although the CDF could be integrated into the same element as the CTF. This would be much like switches work today, with CDR generation integrated within the switch platform. Implementing the CDF as a stand-alone entity raises some interesting opportunities for operators but represents a distinct departure from how billing is accomplished today.

Today, CDRs are created by each network element delivering a service. In the case of a video conference call, the video server would create a CDR for the video portion of the session, while the media gateway controller (MGC) would create the CDR for the voice portion of the call. If there are any test messaging services tied to the conference, then the text server would create a CDR for the text service.

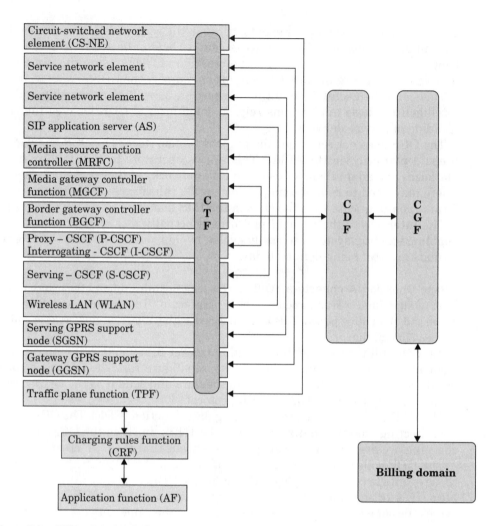

Figure 7.3   Offline charging architecture

By using a more centralized CDF, an operator could deploy the CDR function centrally within a region, for example. This would also enable it to correlate events that are related (tied to one call, for example) and generate one CDR for the entire session.

It also simplifies the delivery of the CDRs, since a common format is chosen for delivery to the billing presentment system. With a centralized CDF function, the CDF performs the mediation we see in today's network. This is necessary because even with standards, vendors tend to create their own implementation of a standard resulting in some proprietary format. This is seen in all networks today.

The formats of the CDRs created by the charging data function (CDF) are defined in 3GPP TS 32.250 (for circuit-switched networks) and 3GPP TS 32.251 (for packet-switched networks). Hopefully we will not see the same drifting from the standards as seen in previous implementations. This only increases cost to deploy new technology and increases cost to applications.

Standardization of the CDRs also allows operators to choose among many different vendors for back-office systems that depend on the CDRs for data. The back-office and operational support systems (OSS/BSS) are crucial to managing a network, and they need standard data formats to be effective.

The CDF then forwards CDRs to the charging gateway function (CGF), which in turn interfaces to the billing system. The charging gateway is responsible for collecting CDRs from multiple CDFs, and for providing a variety of functions, including validation and error handling. CDR storage is also a function of the charging gateway.

This is really where operators can implement CDR management. Storage should be kept to a minimum at the CGF and only used for disaster recovery. Larger storage capacities are best suited for data warehouses in the core network where all data, including usage data and network inventory data, can be stored together for business intelligence and analytics.

The CGF does not process the CDRs but simply manages the delivery of the CDRs under its responsibility. Consider it a collection point within a region or within a network. The charging gateway then manages the transfer of all CDRs to the billing system and can be used to correlate like CDRs, storing them by session type, and even separating by filtering criteria such as originating charging trigger function (CTF). The CGF could be provisioned to forward all messaging CDRs to the billing system at a specific time, for example, and all voice CDRs at another time. This of course is implementation specific but presents some advantages to the operator in the area of billing management.

It also presents an opportunity to forward different CDR types to different billing presentment systems during the migration to IMS, in the event the operator is transitioning multiple billing platforms to support the new technology. This capability would allow the operator to determine where each CDR type is to be sent, and eventually send all CDRs to the same platform once the transition is complete.

Trying to manage these functions at each individual network entity is a monstrous task, requiring configuration at many different systems. This is simply not feasible during large migrations and would only result in more cost to the operator.

The CGF sends CDRs to the billing system either via "push" or "pull" mechanisms, depending on provisioning by the operator. This will depend on the network itself, the billing system and its capabilities, and how the operator wants to manage the transport for billing records. For example the operator may want to let the billing system pull CDRs from the CGF based on a time schedule, providing more control over the receipt of CDRs from multiple CGFs in the network and preventing congestion.

This function, like the CDF can be integrated within the same network element as the CDF, or it may be a stand-alone function. Figure 7.4 illustrates the various options available for implementing these two functions.

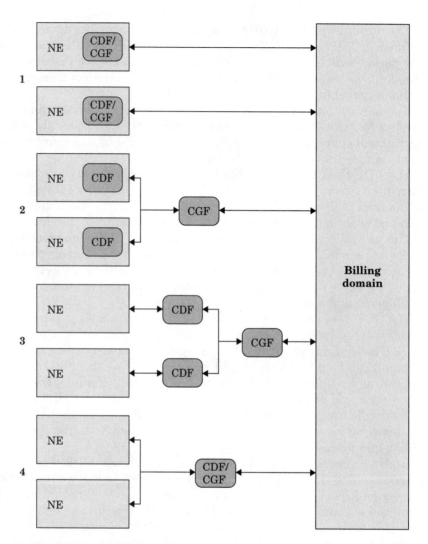

Figure 7.4 Implementation of CDF and CGF

The various implementation options present an opportunity for operators to change the current model used in their billing processes to meet the needs of other processes. One interesting prospect can be seen in option 3 in Figure 7.4, where the CDR creation is a stand-alone function distributed throughout the network.

This allows operators to scale the production of CDRs according to actual volume, and it presents a more efficient method for the management of charging records. Today's model relies on all CDR generation to be resident within each switching element. Should the switching elements become congested, the production of CDRs is secondary to completing a call.

Forwarding the charging event data with DIAMETER to the CGF allows operators more control over their billing data and provides new auditing points that can be monitored for billing verification and auditing. This is explained in more detail toward the end of this chapter.

Option 1 in the figure illustrates the model that exists today, where CDR generation is integrated within the network element itself. This presents a number of issues that we know of today, congestion being one of those problems. When adding other functions in addition to billing (such as various OSS/BSS functions), the use of probes is required to collect the CDRs from the interface to the billing domain. This has already proven extremely expensive in today's legacy networks.

Option 2 could also be an attractive approach, although it still presents the same problem as option 1. With the creation of CDRs integrated within the network element, congestion remains a concern. However, the CGF could provide an interface to the OSS/BSS, eliminating the need for probes in the network.

Option 3 would also allow operators to feed CDRs generated at the CDF through the CGF to other applications in addition to their billing platforms. For example, performance and revenue management solutions, fraud systems, and other OSS/BSS could interface at this point, eliminating the need to place probes all over the network.

Option 4 is much like option 3 in that it provides an attractive option for operators wanting to interface multiple applications to the CGF. One other distinct advantage that both options 3 and 4 present is the ability to store CDRs in a data warehouse. A data warehouse could then be used by various applications to process those CDRs for revenue assurance, fraud management, even security.

There are pros and cons for each of the preceding implementation models. The biggest advantage that centrally located functions provide is the ability to backhaul simple data rather than "formatted" billing records. The data can then be used in a number of different ways, including audits and support of business intelligence systems.

By creating the actual CDRs in the network core, the operator is provided better controls over the actual billing records, not to mention that this increases data integrity, as there are fewer opportunities for CDR corruption when the CDR is not being passed from the edge of the network to the core where the billing system is located.

There is also a greater ability to generate a single CDR for multiple sessions as mentioned before. This is important because some operators may not wish to generate several CDRs for a video conference call (which would require a CDR for each media type being used). By collecting the billing data from each of the network elements at the core, analysis of this data can allow operators to determine how they want to compile the data into a single CDR, and then send that CDR to the billing presentment system at the core.

User Profile

Every subscriber in the IMS must have a user profile. It is the user profile that defines the services a subscriber is authorized to use, and how that subscriber will be charged for those services. The user profile is stored in the IMS core in the home subscriber

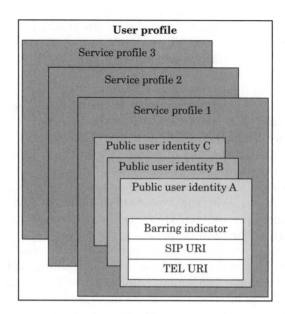

Figure 7.5 Structure of the user profile

server (HSS). This allows operators to create the user profile when a subscriber signs up for service.

The user profile includes the private user identity, its associated public user identities, and the service profiles that are associated with the user. The service profiles will have various public user identities associated with each one, as illustrated in Figure 7.5.

The service profile may also contain the media profile, identifying the SDP parameters that the subscriber is allowed to request from the network. These SDP parameters identify the media resources required to support a session, and will determine what media services the subscriber is allowed to access in the network. There may also be data in the service profile that identifies requests that require support from an application server(s).

The HSS then serves as the central reference point for all charging, since it is the user profile that determines what will be charged. The user profile is then associated with public and private user identities (there can be multiple public user identities associated with one user profile). The private user identity and its associated public user identities are a permanent record within the HSS; however, the location/routing information associated with these identities is not known until the subscriber registers in the network.

When a subscriber registers in the network and is assigned to a Serving Call Session Control Function (S-CSCF), the S-CSCF will query the HSS to acquire the user profile. The user profile will in turn identify the services allowed, the media allowed, and the public user identities associated with that subscriber.

The S-CSCF then uses this information to police the services a subscriber is authorized to request. For example, if the user profile does not authorize messaging services, and the subscriber attempts to initiate messaging through the sending of a *MESSAGE,* the S-CSCF examines the SDP portion of the message, and when it determines that messaging is described in the SDP as the media for the session, the S-CSCF rejects the *INVITE* and sends the appropriate response back to the subscriber.

The use of a user profile provides many opportunities for a service provider. A subscriber may have multiple devices and multiple uses for each of their devices. Some of their uses may be business, while others are strictly pleasure. By providing the ability to define multiple identities under one account, a subscriber can manage all of their communications for both work and personal use using one account, rather than multiple accounts for each identity.

Many wireless providers provide this capability today, but not with the same versatility, because they may not have the ability to support all media types (although this is rapidly changing). Authorizing the services a subscriber is allowed is an important function provided by the IMS.

We will not go into great detail on authorization in this section, since the focus here is in the charging architecture and functions within the IMS specific to charging.

Accounting with DIAMETER

In order for billing systems to be able to accurately tabulate proper charges, billing data must first be collected. The billing data is based on usage measurements for the various services used by the subscriber. Usage measurements can be further defined by the operator according to policy. For example, video services may follow a per-movie fee rather than per-minute fee. Likewise, there may be flat rates for voice calls.

There are two forms of usage measurements used to create charging records. Session-based records are used for voice calls and data transfers. They are measured in terms of duration and bandwidth requirements. This is analogous with how we are billed for telephone calls today (minutes of use in most cases).

This does not mean that all operators must use session-based records. They are there for operators to use if they choose this billing model, but they may choose to use event records instead.

Event-based charging occurs for an event, such as a text message sent or received. These measurements are basically peg counts. This is the simplest form of charging and requires much less processing than session-based charging.

Session-based charging can be difficult in online charging, because the duration of a call is not yet known, so there must be some means of authorizing the call if the subscriber has enough credit in his or her account, yet it has to be a means for terminating the call should the account be depleted. One method specified by 3GPP is the allotment of credit in units.

For example, a unit could be three minutes duration of time. When the call is originated, and the account balance is checked, there could be nine minutes left in the account. Using this principle, the charging function could authorize the call for six minutes, allotting two units for the call.

If the call should use the reserved six minutes, the network element would then create a new charging event, and the billing system would then allot another reservation of units. The new reservation of units would be for the remaining three minutes left on account. If there is no balance on the account, then the call is denied. This is referred to as session-based charging with unit reservation.

Should the call exceed the reserved number of units, the call would be terminated by the network element. Since the network element is in control of the session, it has full control over the bearer path and is able to terminate the session without the use of additional elements or messaging. This is different than legacy prepaid billing systems that use a service node in the bearer path. The service node is then responsible for controlling the termination of the circuit should the account balance be depleted.

Service nodes provide an extra connection in the facilities, forming a form of loop around so that the voice path can be controlled. The alternative would be some form of communications to the switch itself, communicating back to the prepaid billing system. The service node approach is popular in legacy networks because it is vendor agnostic (it works with all switch types). The IMS online charging architecture eliminates the need for service nodes in the call path by placing control back in the network element itself and providing a standardized communications mechanism between the billing system and the network element.

This represents a cost advantage to operators looking to deploy prepaid services. The process is simplified within the IMS model by allowing the network element the means of communicating directly with the charging system regarding an account, and in turn controlling the bearer path.

Session-based charging requires that a CDR be opened as soon as the session begins, and information is added to the CDR throughout the session. This means that interim CDRs may be available but not forwarded to the billing system. The CDF may elect to close a CDR and forward it on based on certain events, such as a volume limit, time limit, or implementation-specific limits.

"Partial" CDRs may be used and made available to other applications in the OSS/BSS for real-time monitoring of a call, depending on how charging is implemented in the network. If a CDR is closed and the session is still in progress, then another CDR will be created for the session. This too would have to be correlated and combined with the other CDRs by the applications in the back office.

The partial CDRs may not contain all of the information for the session. Certainly the first CDR would be a full CDR, containing all of the pertinent information for the session, but subsequent CDRs may only contain changes that occurred during the session, rather than repeating all of the information made available in previous CDRs.

The CGF or the billing system can then be provisioned to accept the partial CDRs and correlate them into one record. This will depend on a number of factors, especially the architecture. If the CDF and the CGF are combined, then it is most likely that the CDRs would have to be correlated and combined by the billing system.

Partial CDRs are often required for real-time applications such as fraud management systems and revenue assurance applications. The purpose of the real-time CDR is to communicate the beginning of a session along with all of the session parameters, and to then report on the progress of the session each time there is a change in the session state.

For example, if another subscriber should be added to the session, this would be communicated in another partial CDR containing the same session ID. The applications can then correlate this CDR with the other, already-received CDRs for the same session and update the status of the session within the application.

Charging in the IMS

The DIAMETER protocol consists of several messages used by the charging entities to exchange charging event data. These messages are shown next. The DIAMETER protocol itself is composed of base attribute-value pairs, or AVPs. Think of AVPs as parameters within the DIAMETER message itself.

- Accounting-Request (ACR)
- Accounting-Answer (ACA)
- Capabilities-Exchange-Request (CER)
- Capabilities-Exchange-Answer (CEA)
- Credit-Control-Request (CCR)
- Credit-Control-Answer (CCA)
- Device-Watchdog-Request (DWR)
- Device-Watchdog-Answer (DWA)
- Re-Auth-Request (RAR)
- Re-Auth-Answer (RAA)

Offline   The *ACR* and *ACA* messages are used for offline charging, while the remainder of these messages are used for online charging. The parameters (or AVPs) for each of these messages can be found in the following tables. Understanding the AVPs for each of these messages helps to understand the information available within the DIAMETER protocol.

The originator of the *ACR* message is the network element (see Table 7.1). Specifically, the CTF within the network entity creates the ACR when a billable event has occurred. What is billable and what is not billable is determined by local policy. The destination of this message is the offline billing system through the CGF. The CGF is responsible for correlating all of the related CDRs and sending them on to the billing system.

TABLE 7.1   Accounting-Request (ACR) Message Content

Parameter	Status
Session-ID	Mandatory, fixed
Origin-host	Mandatory
Origin-realm	Mandatory
Destination-realm	Mandatory
Accounting-record-type	Mandatory
Accounting-record-number	Mandatory
Acct-application-ID	Optional

Vendor-specific-application-ID
    Vendor-ID
    Auth-application-ID (mandatory)
    Acct-application-ID (mandatory)
	Optional

User-name	Optional
Accounting-sub-session-ID	Optional
Accounting-RADIUS-session-ID	Optional
Acct-multi-session-ID	Optional
Acct-interim-interval	Optional
Accounting-realtime-required	Optional
Origin-state-ID	Optional
Event-timestamp	Optional

Proxy-info
    Proxy-host
    Proxy-state
	Optional

Route-record	Optional
AVP	Optional
Event-type	Optional
Role-of-node	Optional
User-session-ID	Optional
Calling-party-address	Optional
Called-party-address	Optional
Time-stamps	Optional

Application-server
    Application-servers-involved
    Application-provided-called-parties
	Optional (IMS S-CSCF only)

Application-provided-called-party-address	Optional (IMS S-CSCF only)

Inter-operator-identifier
    Originating-IOI
    Terminating-IOI
	Optional

IMS-charging-identifier	Optional

TABLE 7.1   Accounting-Request (ACR) Message Content (*continued*)

Parameter	Status
SDP-session-description	Optional

SDP-media-component
    SIP-request-timestamp
    SIP-response-timestamp
    SDP-media-components
        SDP-media-name
        SDP-media-description
        GPRS-charging-ID
    Media-initiator-flag
    Authorized-QoS
	Optional

GGSN-address	Optional
Served-party-IP-address	Optional (IMS P-CSCF only)
Authorized-QOS	Optional (IMS P-CSCF only)
Server-capabilities	Optional (IMS I-CSCF only)
Trunk-group-ID	Optional (IMS MGCF only)
Bearer-service	Optional (IMS MGCF only)
Service-ID	Optional (IMS MRFC only)

UUS-data
    Content-type
    Content-disposition
    Content-length
    Originator
	Optional

Cause	Optional

PS-furnish-charging-information
    GPRS-Charging-ID
    PS-free-format-data
    PS-append-free-format-data
	Optional

The *ACA* message is sent in response to an *ACR* (see Table 7.2). The message is sent from the billing system to the CTF at the network element that originated the request. The CTF within each network element is responsible for creating a chargeable event record and sending it to the forwarding function. In the case of session-based charging, there will be multiple events. There is a start event when the session begins, and a separate event when the session ends. Each of these events is sent to the CDF for correlation and CDR generation.

There may be additional chargeable events if there are changes to a session. For example if a session is changed to add another subscriber, there could be another chargeable event created and forwarded to the CDF. This would eventually be correlated with all of the other chargeable events for the entire session, and correlated into a CDR (or sent as individual CDRs, depending on implementation).

Online   As mentioned earlier, online charging is a prepaid service, where the subscriber establishes an account and an account balance, and that balance is charged each time

TABLE 7.2   Accounting-Answer (ACA) Message Content

Parameter	Status
Session-ID	Mandatory, fixed
Result-code	Mandatory
Origin-host	Mandatory
Origin-realm	Mandatory
Accounting-record-type	Mandatory
Accounting-record-number	Mandatory
Acct-application-ID	Optional

Vendor-specific-application-ID
    Vendor-ID
    Auth-application-ID (mandatory)
    Acct-application-ID (mandatory)
	Optional

User-name	Optional
Accounting-sub-session-ID	Optional
Accounting-RADIUS-session-ID	Optional
Acct-multi-session-ID	Optional
Error-reporting-host	Optional
Acct-interim-interval	Optional
Accounting-realtime-required	Optional
Origin-state-ID	Optional
Event-timestamp	Optional

Proxy-info
    Proxy-host
    Proxy-state
	Optional

AVP	Optional

a service is used. When the account has become depleted, the subscriber must deposit additional monies into the account prior to receiving any additional services.

This means that the network entities must be able to communicate with the prepaid billing system to determine the account balance and ensure services are not delivered when the account has become depleted. The DIAMETER protocol is used for this function.

Credit-control-request is used for online charging (see Table 7.3). The *CCR* message is sent by the network element to the OCS prior to service delivery. This is the message sent to request account balance information based on the service requested by the subscriber. The rating function within the online charging system (OCS) is responsible for providing rate information for each event.

TABLE 7.3 Credit-Control-Request (CCR) Message Content

Parameter	Status
Session-ID	Mandatory, fixed
Origin-host	Mandatory
Origin-realm	Mandatory
Destination-realm	Mandatory
Auth-application-ID	Mandatory
Destination-host	Mandatory

Vendor-specific-application-ID
  Vendor-ID
  Auth-application-ID (mandatory)
  Acct-application-ID (mandatory)
	Optional

User-name	Optional
Acct-multi-session-ID	Optional
Origin-state-ID	Optional
Event-timestamp	Optional

Proxy-info
  Proxy-host (mandatory)
  Proxy-state (mandatory)
	Optional

Route-record	Optional
Termination-cause	Optional
AVP	Optional
CC-request-type	Mandatory
CC-request-number	Mandatory
CC-subsession-ID	Mandatory

Subscription-ID
  Subscription-ID-type (mandatory)
  Subscription-ID-data (mandatory)
	Optional

Requested-action	Optional

Requested-service-unit
  CC-time
  CC-money
    Unit-value (mandatory)
    Value-digits (mandatory)
    Exponent
    Currency-code
  CC-total-octets
  CC-input-octets
  CC-output-octets
  CC-service-specific-units
  AVP
	Optional

(Continued)

TABLE 7.3   Credit-Control-Request (CCR) Message Content (*continued*)

Parameter	Status
Used-service-unit	Optional
Tariff-change-usage	
CC-time	
CC-money	
Unit-value (mandatory)	
Value-digits (mandatory)	
Exponent	
Currency-code	
CC-total-octets	
CC-input-octets	
CC-output-octets	
CC-service-specific-units	
AVP	
Service-parameter-info	Optional
Service-parameter-type	
Service-parameter-value	
CC-correlation-ID	Optional
Service-identifier	Optional
Multiple-services-indicator	Optional
Multiple-services-credit-control	Optional
Reporting-reason	
Trigger-type	
Granted-service-unit	
Requested-service-unit	
CC-time	
CC-money	
Unit-value (mandatory)	
Value-digits (mandatory)	
Exponent	
Currency-code	
CC-total-octets	
CC-input-octets	
CC-output-octets	
CC-service-specific-units	
AVP	
Used-service-unit	
Reporting-reason	
Tariff-change-usage	
CC-time	
CC-money	
Unit-value (mandatory)	
Value-digits (mandatory)	
Exponent	
Currency-code	
CC-total-octets	
CC-input-octets	
CC-output-octets	
CC-service-specific-units	
AVP	

TABLE 7.3   Credit-Control-Request (CCR) Message Content (*continued*)

Parameter	Status
Tariff-change-usage Service-identifier Rating-group G-S-U-pool-reference Validity-time Result-code Final-unit-indication AVP	

User-equipment-info
  User-equipment-info-type (mandatory)
  User-equipment-info-value (mandatory)
	Optional

Service-information
  PS-information
  WLAN-information
  IMS-information
  MMS-information
  LCS-information
	Optional

Credit-control-answer (CCA) is also used for online charging and is sent by the OCS to the network element in response to a *CCR* (see Table 7.4). The response will carry with it the amount of credit being authorized for the session. The amount of credit is based on the available account balance in the prepaid system. The credit is provided in increments based on local policy, and the amount of credit provided in the *Result-code* parameter.

TABLE 7.4   Credit-Control-Answer (CCA) Message Content

Parameter	Status
Session-ID	Mandatory, fixed
Result-code	Mandatory
Origin-host	Mandatory
Origin-realm	Mandatory
Auth-application-ID	Mandatory

Vendor-specific-application-ID
  Vendor-ID
  Auth-application-ID (mandatory)
  Acct-application-ID (mandatory)
	Optional

User-name	Optional
Acct-multi-session-ID	Optional
Redirect-host	Optional
Redirect-host-usage	Optional
Redirect-max-cache-time	Optional
Origin-state-ID	Optional

(Continued)

TABLE 7.4 **Credit-Control-Answer (CCA) Message Content (*continued*)**

Parameter	Status
Event-timestamp	Optional
Proxy-info Proxy-host (mandatory) Proxy-state (mandatory)	Optional
Route-record	Optional
AVP	Optional
CC-request-type	Mandatory
CC-request-number	Mandatory
CC-subsession-ID	Optional
CC-session-failover	Optional
Subscription-ID	Optional
Granted-service-unit Tariff-time-change CC-time Unit-value (mandatory) Value-digits (mandatory) Exponent Currency-code CC-total-octets CC-input-octets CC-output-octets CC-service-specific-units Time-quota-threshold Volume-quota-threshold AVP	Optional
Cost-information Unit-value (mandatory) Value-digits (mandatory) Exponent Currency-code (mandatory) Cost-unit	Optional
Final-unit-indication Final-unit-action (mandatory) Restriction-filter-rule Filter-ID Redirect-server	Optional
Check-balance-result	Optional
Credit-control-failure-handling	Optional
Validity-time	Optional
Trigger-type	Optional
Direct-debiting-failure-handling	Optional

TABLE 7.4   Credit-Control-Answer (CCA) Message Content (*continued*)

Parameter	Status
Multiple-services-credit-control	Optional
Quota-holding-time	
Granted-service-unit	
Tariff-time-change	
CC-time	
CC-money	
Unit-value (mandatory)	
Value-digits (mandatory)	
Exponent	
Currency-code	
CC-total-octets	
CC-input-octets	
CC-output-octets	
CC-service-specific-units	
Time-quota-threshold	
Volume-quota-threshold	
AVP	
Requested-service-unit	
Used-service-unit	
Tariff-change-usage	
Service-identifier	
Rating-group	
G-S-U-Pool-reference	
G-S-U-pool-identifier (mandatory)	
CC-unit-type (mandatory)	
Unit-value (mandatory)	
Validity-time	
Result-code	
Final-unit-indication	
Final-unit-action	
Restriction-filter-rule	
Filter-ID	
Redirect-server	
Redirect-address-type (mandatory)	
Redirect-server-address (mandatory)	
AVP	
PS-furnish-charging-information	
GPRS-charging-ID (mandatory)	
PS-free-format-data (mandatory)	
PS-append-free-format-data	

Re-auth-request (RAR) is used for online charging and is sent by the OCS to the network element (see Table 7.5). This is used when the original credit has been exhausted but the session is going to continue. The network element is able to request additional credits for the same session.

Re-auth-answer (RAA) is sent by the network back to the OCS in response to an *RAR* (see Table 7.6). This response will identify the amount of credit being given for the continuing session. If the balance on the account has been depleted, then the request for additional credit is denied in this response.

TABLE 7.5   Re-Auth-Request (RAR) Message Content

Parameter	Status
Session-ID	Mandatory, fixed
Origin-host	Mandatory
Origin-realm	Mandatory
Destination-realm	Mandatory
Destination-host	Mandatory
Auth-application-ID	Mandatory
Re-auth-request-type	Mandatory
User-name	Optional
Origin-state-ID	Optional
Event-timestamp	Optional

Proxy-info
   Proxy-host (mandatory)
   Proxy-state (mandatory)
	Optional

Route-record	Optional
AVP	Optional
CC-sub-session-ID	Optional
G-S-U-Pool-Identifier	Optional
Service-identifier	Optional
Rating-group	Optional

TABLE 7.6   Re-Auth-Answer (RAA) Content Message

Parameter	Status
Session-ID	Mandatory, fixed
Result-code	Mandatory
Origin-host	Mandatory
Origin-realm	Mandatory
User-name	Optional
Origin-state-ID	Optional
Error-message	Optional
Error-reporting-host	Optional
Failed-AVP	Optional
Redirect-host	Optional
Redirect-host-usage	Optional
Redirect-host-cache-time	Optional

Proxy-info
   Proxy-host (mandatory)
   Proxy-state (mandatory)
	Optional

AVP	Optional

TABLE 7.7   Capabilities-Exchange-Request (CER) Message Content

Parameter	Status
Origin-host	Mandatory
Origin-realm	Mandatory
Host-IP-address	Mandatory
Vendor-ID	Mandatory
Product-name	Mandatory
Origin-state-ID	Mandatory
Supported-vendor-ID	Optional
Auth-application-ID	Optional
Inband-security-ID	Optional
Acct-application-ID	Optional
Vendor-specific-application-ID	Optional
Firmware-revision	Optional
AVP	Optional

The *Capabilities-exchange-request (CER)* message can be sent by either the network element or the OCS (see Table 7.7). It is used between two DIAMETER servers to exchange information about the capabilities of the server. Either end of a DIAMETER dialog would send this to the other node to request information about the capabilities of the server.

The *Capabilities-exchange-answer (CEA)* message is sent in response to the *CER,* by either the network element or the OCS (depending on who originated the *CER*; see Table 7.8). The answer provided in this message communicates the capabilities of the server or entity answering. For example, a charging platform would use this message to communicate to a network element that it supports the DIAMETER protocol and supports online charging.

Device-watchdog-request (DWR) is sent by either the network element or the OCS, when there has been a lapse in traffic between the two nodes (see Table 7.9). The request is sent to determine if there was a failure in the facilities. The sending node is expecting to receive an answer in response; if no response (DWA) is received, then the node assumes that the communications path between the two entities has failed.

Device-watchdog-answer (DWA) is sent in response to a *DWR,* by either the network element or the OCS (dependent on who originated the *DWR*; see Table 7.10). The DWA assures the opposite node that transmission between the two entities is still operational.

DIAMETER Attribute-Value Pairs (AVPs)   The AVPs are the parameters used within the DIAMETER messages to provide information regarding an event or the originator of an event. The AVPs for each message were identified in the preceding section. This section identifies the additional details of the AVPs.

An AVP can consist of several other AVPs grouped together. These are referred to as grouped AVPs. There are baselines of AVPs that are defined in the DIAMETER RFC,

TABLE 7.8   Capabilities-Exchange-Answer (CEA) Message Content

Parameter	Status
Result-code	Mandatory
Origin-host	Mandatory
Origin-realm	Mandatory
Host-IP-address	Mandatory
Vendor-ID	Mandatory
Product-name	Mandatory
Origin-state-ID	Optional
Error-message	Optional
Failed-AVP	Optional
Supported-vendor-ID	Optional
Auth-application-ID	Optional
Inband-security-ID	Optional
Acct-application-ID	Optional
Vendor-specific-application-ID	Optional
Firmware-revision	Optional
AVP	Optional

TABLE 7.9   Device-Watchdog-Request (DWR) Message Content

Parameter	Status
Origin-host	Mandatory
Origin-realm	Mandatory
Origin-state-ID	Mandatory

TABLE 7.10   Device-Watchdog-Answer (DWA) Message Content

Parameter	Status
Result-code	Mandatory
Origin-host	Mandatory
Origin-realm	Mandatory
Error-message	Optional
Failed-AVP	Optional
Original-state-ID	Optional

and additional AVPs defined specifically for the use in an IMS domain. Some AVPs are specific to offline charging, while others are specific to online charging.

The previous sections defined the DIAMETER messages specific to online and offline charging respectively. This section defines the AVPs used within those messages for both online and offline (as indicated). All AVPs are provided in alphabetical order for ease of searching, but Table 7.11 identifies the AVPs and how they are used. One additional note: an AVP is really a parameter within one of the DIAMETER messages, so you will see this terminology used interchangeably with AVP.

TABLE 7.11   DIAMETER AVPs

AVP Name	AVP Code*	Use
Acct-application-ID	259	Baseline
Amount-of-UUS-data	857	Offline
Application-provided-called-party-address	837	Offline
Application-server	836	Offline
Authorized-QOS	849	Offline
Bearer-service	854	Offline
Called-party-address	832	Offline
Calling-party-address	831	Offline
Cause	860	Offline
Cause-code	861	Offline
CC-correlation-ID	TBD	Online
CC-input-octets	TBD	Online
CC-money	TBD	Online
CC-output-octets	TBD	Online
CC-request-number	TBD	Online
CC-request-type	TBD	Online
CC-service-specific-units	TBD	Online
CC-session-failover	TBD	Online
CC-sub-session-ID	TBD	Online
CC-time	TBD	Online
CC-total-octets	TBD	Online
CC-unit-type	TBD	Online
Check-balance-result	TBD	Online
Content-disposition	828	Offline
Content-length	827	Offline
Content-type	826	Offline
Cost-information	TBD	Online
Cost-unit	TBD	Online

(Continued)

TABLE 7.11   DIAMETER AVPs (*continued*)

AVP Name	AVP Code*	Use
Credit-control	TBD	Online
Credit-control-failure-handling	TBD	Online
Currency-code	TBD	Online
Direct-debiting-failure-handling	TBD	Online
Direction	859	Offline
Event	825	Offline
Event-type	823	Offline
Exponent	TBD	Online
Final-unit-action	TBD	Online
Final-unit-indication	TBD	Online
GGSN-address	847	Offline
GPRS-charging-ID	846	Offline
GPRS-charging-ID	846	Online
Granted-service-unit	TBD	Online
Granted-service-unit-pool-identifier	TBD	Online
Granted-service-unit-pool-reference	TBD	Online
IMS-charging-identifier	841	Offline
Incoming-trunk-group-ID	852	Offline
Inter-operator-identifier	838	Offline
Mime-type	858	Offline
Multiple-service-credit-control	TBD	Online
Multiple-service-indicator	TBD	Online
Node-functionality	862	Offline
Originating-IOI	839	Offline
Outgoing-trunk-group-ID	853	Offline
PS-append-free-format-data	867	Online
PS-free-format-data	866	Online
PS-furnish-charging-information	865	Online
Quota-holding-time	871	Online
Rating-group	TBD	Online
Redirect-address-type	TBD	Online
Redirect-server	TBD	Online
Redirect-server-address	TBD	Online
Reporting-reason	872	Online
Requested-action	TBD	Online
Requested-service-agent	TBD	Online
Restriction-filter-rule	TBD	Online
Result-code	298	Baseline
Role-of-node	829	Offline

TABLE 7.11   DIAMETER AVPs (*continued*)

AVP Name	AVP Code*	Use
SDP-media-component	843	Offline
SDP-media-description	845	Offline
SDP-media-name	844	Offline
SDP-session-description	842	Offline
Served-party-IP-address	848	Offline
Server-capabilities	TBD	Offline
Service-ID	855	Offline
Service-identifier	TBD	Online
Service-information	TBD	Online
Service-parameter-info	TBD	Online
Service-parameter-type	TBD	Online
Service-parameter-value	TBD	Online
SIP-method	824	Offline
SIP-request-timestamp	834	Offline
SIP-response-timestamp	835	Offline
Subscription-ID	TBD	Online
Subscription-ID-data	TBD	Online
Subscription-ID-type	TBD	Online
Tariff-change-usage	TBD	Online
Tariff-time-change	TBD	Online
Terminating-IOI	840	Offline
Time-quota-threshold	868	Online
Time-stamps	833	Offline
Trigger-type	870	Online
Trunk-group-ID	851	Offline
Unit-value	TBD	Online
Used-service-unit	TBD	Online
User-equipment-info	TBD	Online
User-equipment-info-type	TBD	Online
User-equipment-info-value	TBD	Online
User-name	1	Baseline
User-session-ID	830	Offline
UUS-data	856	Offline
Validity-time	TBD	Online
Value-digits	TBD	Online
Vendor-ID	266	Baseline
Volume-quota-threshold	869	Online

* The TBD in the AVP Code column indicates that the RFC defining these AVPs is still in draft and the codes have not yet been defined.

There are several AVPs that are not yet defined by the RFCs. This is still work in progress at the time this book went to print. If there is no definition given for an AVP, and no code assigned in Table 7.11, the committee is still working on fully defining the AVP and its use. They will be defined in future revisions of this book.

Acct-Application-ID   This parameter is found in the *Vendor-specific-application-ID* message; it contains the value of 1. This AVP is only used in offline charging within the IMS domain.

Amount-of-UUS-data   The amount of user-to-user data is identified in this AVP. The value identifies the number of octets worth of data carried in the message body of a SIP message. This could then be used by the operator to charge based on data size (the amount of user data received/sent).

Application-provided-called-party-address   The called party number in the form of a SIP URI or E.164 number is provided in this AVP if the application server is able to determine what the called party number is. This information can be useful in determining if the called number was different than the number being routed to (for example, the called number was a subscriber's work number, but the session was redirected to the subscriber's home number).

Application-server   This AVP provides the URL of the application server that is used for the identified session. There may be multiple application servers defined for service delivery.

Authorized-QoS   This AVP identifies the QoS applied on the Go interface if applicable.

Bearer-service   This AVP identifies the bearer service provided on the analog (PSTN) side of the session. The bearer service may be voice, for example.

Called-party-address   The called party address identifies the subscription to which a session was established. The address can be in the form of a SIP URI, or an E.164 number.

Calling-party-address   The calling party address identifies the subscription that originated a session. This address can also be in the form of a SIP URI, or an E.164 number.

Cause   This AVP consists of the *Cause-code,* and the *Node-functionality* parameters. The *Node-functionality* identifies the node that generated the *Cause-code.* The *Cause-code* identifies whether or not an accounting function was successful or if it failed, and if it failed, provides a reason for the failure.

Cause-code   The cause code identifies whether or not an accounting function was successful, and if not, what caused the failure. The following values are used:

- *Normal end of session*   Indicates that the SIP session was ended normally via a SIP BYE message (and not by error).
- *Successful transaction*   Indicates a successful SIP transaction (*Register, Invite,* etc.).
- *End of SUBSCRIBE dialog*   Indicates the closure of a *SUBSCRIBE* dialog.
- *3xx redirection*   Indicates a transaction was terminated due to a 3*xx* response.
- *Unspecified error*   Indicates the SIP transaction was terminated due to some error, but the error cannot be identified.
- *4xx response failure*   Indicates the SIP transaction was terminated because an IMS node received a 4*xx* response.
- *5xx server failure*   Indicates the SIP transaction was terminated because an IMS node received a 5*xx* response.
- *6xx global failure*   Indicates the SIP transaction was terminated because an IMS node received a 6*xx* response.
- *Unsuccessful session setup*   Indicates the session was not set up successfully.
- *Internal error*   Indicates the session was terminated due to an internal error in the IMS node.

Content-disposition   Defines how a SIP message body is to be interpreted. For example, the message body identifies the parameters of a session, or the message body contains a text message.

Content-length   Identifies the size of the SIP message body, if user data is being exchanged in the SIP message body. This information can then be used to determine billing for data such as text messaging.

Content-type   Identifies whether the message body is text or contains an SDP, for example.

Direction   Identifies the direction of the user-to-user data, with two possible values:

- Uplink = 0
- Downlink = 1

Use of this information will vary, but it can be applied to rate plans. For example, one rate can be applied for uplink data, while a different rate plan is applied to downlink data.

Event   This parameter holds the content of the *EVENT* header found in the *SUBSCRIBE* and *NOTIFY* message received by the network entity. The data is removed from the SIP message and provided in the DIAMETER message.

Event-type   Found in the *Accounting-event* message, *Event-type* identifies the type of service for which the request was generated. There are several parameters to *Event-type*:

- *SIP-method*
- *Event*
- *Content-type*
- *Content-length*
- *Content-disposition*

GGSN-address   This parameter identifies the GGSN that generated the charging ID located in another AVP in the same message. The GGSN-address is in the form of the GGSN's IP address.

GPRS-charging-ID   The GGSN provides a sequence number in this parameter at the time the PDP context is created.

IMS-charging-identifier (ICID)   This parameter identifies the IMS charging identifier for a SIP session. It is generated by an IMS node and derived from SIP messaging. The purpose of the ICID is to provide an identity that receiving nodes can use for correlation of multiple associated charging messages.

Incoming-trunk-group-ID   The incoming trunk identified in this parameter identifies the trunk circuit from the PSTN portion of the associated call.

Inter-operator-identifier   When a call is transported between two operators, inter-operator identification must be provided for proper inter-carrier billing. The identification of the origination and the terminating operators are exchanged in the SIP signaling, and extracted from SIP for inclusion in this parameter. The parameter consists of two values:

- *Originating inter-operator identifier (IOI)*
- *Terminating inter-operator identifier (IOI)*

Mime-type   When user-to-user data is being exchanged within the SIP message body and is in the MIME format, this parameter identifies the MIME type used for the data.

Node-functionality   This AVP is used to identify the function of the node that generated a cause code. Each IMS node type is identified by a specific number as follows:

- *0* = S-CSCF
- *1* = P-CSCF
- *2* = I-CSCF
- *3* = Media Resource Function Controller (MRFC)
- *4* = Media Gateway Controller Function (MGCF)
- *5* = Border Gateway Control Function (BGCF)
- *6* = Application server (AS)
- *7* = User equipment (UE)

Originating-IOI   The originating inter-operator identifier in this parameter is created by the S-CSCF in the home network of the subscriber who originated the session.

Outgoing-trunk-group-ID   This parameter identifies the trunk group in the PSTN for the outgoing portion of the associated call.

PS-append-free-format-data   This AVP is used to indicate to the GGSN whether or not it needs to "append" or overwrite stored data from a previous *PS-free-format-data* AVP. If there is no data stored on the GGSN, then the AVP is ignored. This AVP allows for additional data about a session to be sent, or have that stored data replaced with new session data. There are two values for this parameter:

- *0* = *Append*
- *1* = *Overwrite*

PS-free-format-data   This parameter provides session data specific to an online charging.

PS-furnish-charging-information   When CDRs are being generated for prepaid service (concurrent with online charging), the session information that is needed for those CDRs is provided within the AVP. This information is usually found in a CCA message and sent via the Ro interface (online charging). However, it can also be sent transparently over the Rf interface if online charging and CDR generation are running in parallel. This AVP is a grouping of the following AVPs:

- *GPRS-charging-ID*
- *PS-free-format-data*
- *PS-append-free-format-data*

Quota-holding-time   If a quota (for time or octets) has been granted, and transmission stops, then the device begins a timer. The timer signifies the end of traffic, so that "dead air" is not charged against the balance of the account. The holding time is the amount of time that the device is to hold before it makes the assumption that the quota has expired.

Reporting-reason   This parameter identifies why reporting has occurred, based on the reasons provided. The values for this parameter are

- *Threshold*   Usage reporting occurred because the threshold has been reached.
- *QHT*   The quota holding time has been met, meaning that no usage occurred for the amount of time specified in the *Quota-holding-time* parameter.
- *Final*   Indicates a normal termination of the PDP context.
- *Quota_exhausted*   Indicates that the quota has been exhausted.
- *Validity_time*   Indicates that the credit authorization time limit has expired.
- *Other_quota_type*   If a multidimensional quota is in use, then there was a trigger that occurred and the other quota type is being reported.
- *Rating_condition_change*   A change has occurred in rating conditions. The event that caused the rating condition to change is identified in a subsequent *Trigger-type* AVP.
- *Forced_reauthorization*   A RAR has been received and reauthorization is being forced by a server.

Result-code   Result codes are used to identify why an *Accounting-request* may have failed. The *result-code* is only used for offline charging within an IMS domain. The following values are used in this AVP:

- *Diameter_end_user_service_denied*   4010: This is sent by the OCS when the requested service cannot be delivered because either the subscriber is not allowed to access the service or the account balance is not enough to cover the service.
- *Diameter_credit_control_not_applicable*   4011: The OCS returns this AVP to indicate that there is no credit control required, because there is no charge for the service requested.
- *Diameter_credit_limit_reached*   4012: This AVP indicates that the subscriber's account has reached its limit; therefore, the requested service cannot be delivered.
- *Diameter_user_unknown*   5030: The subscriber identified in the service request cannot be found in the OCS.
- *Diameter_rating_failed*   5031: This code indicates that rating was not possible because the information provided to the rating function was either not valid or incomplete; therefore, the requested rating could not be completed.

Role-of-node   When an application server (AS) or a CSCF identifies itself and its functions for the purposes of charging, it also identifies its role in the associated session. The values for this AVP are

- *0 = Originating role*
- *1 = Terminating role*
- *2 = Proxy role (application servers only)*
- *3 = B2BUA (application servers only)*

SDP-media-component   The media used for an IMS session are described in this AVP. The values can be

- *SDP-media-name*
- *SDP-media-description*
- *GPRS-charging-ID*

SDP-media-description   This parameter copies the media description line from the SDP section of a SIP message. The media description lines describe the type of media that the session is using. It is carried here as part of the charging function for the purposes of billing to describe the media type used in the associated session. The attribute lines are

- *i = media title*
- *c = connection information*
- *b = bandwidth information*
- *k = encryption key*
- *a = zero or more attribute lines*

SDP-media-name   This AVP is also derived from the SDP portion of a SIP message and contains the contents of the "*m=media name*" line.

SDP-session-description   Like the *SDP-media-description,* this AVP also contains the contents of an attribute line from the SDP portion of a SIP message. While the *SDP-media-description* identifies the type of media, this is used to describe the type of session.

Served-party-IP-address   If the P-CSCF is providing service to the calling or the called party, then the IP address of either party (or both parties) is provided here.

Service-ID   The media resource function controller (MRFC) fills this AVP identifying the type of service that is being provided (i.e., conference bridge).

Service-information   The format for this AVP varies depending on the specific service being identified. There are several different types of information carried within the specific parameters of this AVP, providing information about the session. The other AVPs within this group and contained in the *service-information* AVP are

- *PS-information*
- *WLAN-information*
- *IMS-information*
- *MMS-information*
- *LCS-information*

SIP-method   This parameter identifies the SIP method (*INVITE, UPDATE,* etc.) that caused an accounting request to be made.

SIP-request-timestamp   This parameter provides the timestamp from the initial SIP request that was used to start the session associated with this charging event.

SIP-response-timestamp   This parameter provides the timestamp from the SIP response used to respond to the initial request (i.e., a *200 OK* sent in response to an *INVITE*).

Terminating-IOI   Identifies the terminating operator for a session, and is generated by the S-CSCF in the home network of the called subscriber (terminating end of the session).

Time-quota-threshold   This AVP is used whenever there was a *Multiple-services-credit-control* AVP received containing a *Granted-service-units* AVP with a time value in the *CC-time* parameter. This value identifies the threshold for the assigned time quota. When the threshold is passed, the network element will then send a reauthorization to the credit control function to have another number of time units provided. The session will remain in progress until the results are provided back with a new time quota.

Time-stamps   This AVP consists of two timestamps, both of which have already been described. They are

- *SIP-request-timestamp*
- *SIP-response-timestamp*

Trigger-type   The events listed in this parameter identify the type of events that must occur to cause a reauthorization to take place. These events will then trigger a request for additional time or octets allowed based on the account status. There are several events identified in this parameter:

- *Change_in_SGSN_IP_address*
- *Change_in_QoS*

- *Change_in_location*
- *Change_in_RAT (radio access technology)*

Trunk-group-ID   This AVP consists of two trunk group identifiers, the outgoing and the incoming. They are identified in the following AVPs:

- *Incoming-trunk-group-ID*
- *Outgoing-trunk-group-ID*

User-name AVP   The user name in this case is the private user identity provided by the network element if it is available. It is only used for offline charging in the IMS domain.

User-session-ID   The session ID in a SIP session is the *CALL ID* derived from the SIP message.

UUS-data   This contains information about user-to-user data that was sent; it consists of the following values:

- *Amount-of-UUS-data*
- *Mime-type*
- *Direction*

Vendor-ID   In the case of IMS, the vendor ID carries a value of *10415,* indicating "3GPP." This AVP is found in the *Vendor-specific-application-ID* message. This AVP is only used for offline charging in an IMS domain.

Volume-quota-threshold   Like the *Time-quota-threshold,* this AVP is used to identify the number of octets allotted for a session under credit control (prepaid). When the quota threshold has been exceeded, the network element must query the OCS for additional authorization (and an additional quota) or terminate the session. The threshold identifies when the reauthorization must take place.

Online Charging Procedures and Message Flow   There are several messages needed for online charging as already outlined, *Credit-control-request* and *Credit-control-answer* being the primary messages. When a subscriber requests service, the serving node sends a *Credit-control-request* to the OCS. The OCS bears the responsibility of rating the event (through the rating function) and then verifying the account balance of the subscriber (the account balance management function). If there is no balance in the account, of course the event is denied and the OCS sends the appropriate message back to the CTF.

If there is a balance in the account, the amount needed for the event is established through the rating function. The amount is then sent to the ABMF to be debited from the account.

Event sessions are handled somewhat differently, since there is no time base involved. For example, a subscriber elects to subscribe to a weather alert service from their local operator. Their device uses the *SUBSCRIBE* method for establishing the subscription to the service in the SIP domain.

The subscription request is sent to the application server (AS) providing the weather service after the event has been rated by the rating function. The AS then sends a DIAMETER *Credit-control-request* message to the OCS to determine how to rate for the event. The OCS sends the *Credit-control-answer* back to the AS so that the service can begin. The ABMF function debits the subscriber's account immediately.

If this is a session-based event, then the amount needed is not yet known. This may then require "unit reservation." This is where the exact usage may not be known so a reserve amount is provided.

When the reserve amount is depleted, the serving node sends another *Credit-control-request,* or a *Re-auth-request* to the OCS, which is then rated, and another reserve amount allocated. If the session is completed and there is excess in the reserved amount allocated, then the excess can be credited back to the subscriber account.

The OCS is ultimately responsible for approving a session based on available credit or denying service. There are provisions that would allow the OCS to grant services even if an account were depleted, leaving the decision process to an operator's specific implementation. This means that there is no absolute rule that says when an account is depleted, service will be denied.

The network element itself has the ultimate responsibility to manage the session. Given the number of units allowed from the OCS, the network element is responsible for terminating a call once the units have been depleted (the account balance then is depleted). The network element can then launch another charging event to the OCS for additional units for the call to continue.

Table 7.12 identifies the various triggers within SIP or ISUP that would cause a network entity to send a DIAMETER message to the OCS for charging.

Offline Charging Procedures and Message Flow   Offline charging requires the generation of CDRs prior to charging for an event. Unlike online charging, offline charging is a historical event. The generated CDRs are collected and stored for a billing period. Service is delivered upon request.

This means there is a lot more messaging involved to control the generation and storage of these CDRs, and to indicate when generation should start and stop. For example, when a session begins, the serving entity may send a *Credit-Control-Request (CCR)* to begin the creation of a CDR. The CDR cannot be used to determine duration, however, so a subsequent CCR would have to be sent once the session ended. The CGF has the responsibility of correlating all of the CDRs generated for one session and sending them all as one CDR.

It is possible then to see several CCRs and CCAs sent between the network elements and the offline charging system to establish CDRs for a session. The CDRs are then sent by the CGF to a billing system where they can be rated and sent to invoicing for presentment.

The Table 7.13 identifies the various triggers that would generate the CCR message.

TABLE 7.12   CCR Triggers

DIAMETER Message	SIP Trigger
CCR [initial]	SIP INVITE
	SIP NOTIFY
	SIP MESSAGE
	SIP REGISTER
	SIP SUBSCRIBE
	SIP REFER
	SIP PUBLISH
CCR [update]	SIP 200 OK (acknowledging *INVITE*, or *UPDATE*)
CCR [terminate]	SIP BYE
	SIP 200 OK (acknowledging non-session SIP methods)
	SIP CANCEL
	SIP 2*xx* (except 200 OK)
	SIP 3*xx*, 4*xx*, 5*xx*, 6*xx*
CCR [event]	SIP NOTIFY
	SIP MESSAGE
	SIP REGISTER
	SIP REFER
	SIP PUBLISH
	SIP 4*xx*, 5*xx*, 6*xx*

TABLE 7.13   ACR Triggers

DIAMETER Message	SIP/ISUP Trigger
ACR [start]	SIP 200 OK (acknowledging receipt of an initial *INVITE*)
	ISUP ANM
ACR [interim]	SIP 200 OK (acknowledging receipt of a subsequent *INVITE* or *UPDATE*)
	Expiration of AVP
ACR [stop]	SIP BYE
	ISUP REL
ACR [event]	SIP 200 OK (acknowledging *NOTIFY, MESSAGE, REGISTER, SUBSCRIBE*, or *PUBLISH*)
	SIP 202 ACCEPTED (acknowledging REFER)
	SIP 2*xx* (not 200 OK)
	SIP 3*xx*, 4*xx*, 5*xx*, 6*xx* Final Response
	SIP CANCEL

Charging Data Correlation   When multiple network entities are involved with the same session, there will be multiple CDRs from all of these different elements. The SIP protocol is used to correlate all of the CDRs received from the various entities. Specifically, the IMS Charging Identifier (ICID) parameter is sent in a *P-CHARGING-VECTOR* header from the origination to the termination point and included in all CDRs for the same session.

The *ICID* is created by the first IMS entity to receive a request. For example, if a call originated in the PSTN and was transiting the IMS domain, the P-CSCF may be the first to receive the request (i.e., an *INVITE*) and would create the unique *ICID* for the session. This *ICID* is then used in the response and all other requests/responses associated with the session until the session is terminated.

When a session originates outside of the IMS, say in the GPRS domain, another parameter is used to correlate CDRs from that domain. The *ACCESS NETWORK CHARGING IDENTIFIER* parameter, found in the *P-CHARGING-VECTOR* header, provides another unique identifier from the access network.

For example, in the case of GPRS the identifier would be a combination of the GGSN address and the PDP context identifier. This identifier would then be used along with the *ICID* to correlate CDRs from end to end.

Appendix

A

3GPP Documentation

IMS-Related Documentation

TS21.133 Security Threats and Requirements

TS21.905 3G Vocabulary

TS22.002 Circuit Bearer Services (BS) Supported by a Public Land Mobile Network (PLMN)

TS22.003 Circuit Teleservices Supported by a Public Land Mobile Network (PLMN)

TS22.004 General on Supplementary Services

TS22.016 International Mobile Station Equipment Identities (IMEI)

TS22.022 Personalization of Mobile Equipment (ME); Mobile functionality specification

TS22.024 Description of Charge Advice Information (CAI)

TS22.041 Operator Determined Barring (ODB)

TS22.071 Location Services (LCS); Service Description

TS22.086 Advice of Charge (AoC) Supplementary Services

TS22.105 Services and Service Capabilities

TS22.115 Service Aspects; Charging and Billing

TS22.127 Open Service Access (OSA)

TS22.140 Multimedia Messaging Service

TS22.141 Presence Service

TS22.228 Service Requirements for the IP Multimedia Core Network

TS22.250 IMS Group Management

TS22.340 IMS Messaging; Stage 1

TS22.931 IP Based Multimedia Services Framework

TS22.940 IP Multimedia Subsystem (IMS) Messaging

TS23.002 Network Architecture

TS23.003 Numbering, Addressing, and Identification

TS23.008 Organization of Subscription Data

TS23.009 Handover Procedures

TS23.011 Technical Realization of Supplementary Services

TS23.012 Location Management Procedures

TS23.015 Technical Realization of Operator Determined Barring (ODB)

TS23.032 Universal Geographical Area Description (GAD)

TS23.040 Technical Realization of the Short Message Service (SMS)

TS23.041 Technical Realization of Cell Broadcast Service (CBS)

TS23.042 Compression Algorithm for Text Messaging Services

TS23.048 Security Mechanisms for the (U)SIM Application Toolkit

TS23.060 General Packet Radio Service (GPRS); Service Description; Stage 2

TS23.066 Support of Mobile Number Portability (MNP); Technical Realization

TS23.072 Call Deflection Supplementary Service

TS23.078 Customized Applications for Mobile Network Enhanced Logic (CAMEL)

TS23.079 Support of Optimal Routing (SOR); Technical Realization

TS23.081 Line Identification Supplementary Services

TS23.082 Call Forwarding (CF) Supplementary Services

TS23.083 Call Waiting (CW) and Call Hold (HOLD) Supplementary Service

TS23.084 Multi Party (MTPY) Supplementary Service

TS23.085 Closed User Group (CUG) Supplementary Service

TS23.086 Advice of Charge (AoC) Supplementary Services

TS23.087 User-to-User Signalling (UUS) Supplementary Service

TS23.088 Call Barring (CB) Supplementary Service

TS23.091 Explicit Call Transfer (ECT) Supplementary Service

TS23.093 Technical Realization of Completion of Calls to Busy Subscriber (CCBS)

TS23.097 Multiple Subscriber Profile (MSP)

TS23.101 General UMTS Architecture

TS23.107 Quality of Service (QoS) Concept and Architecture

TS23.125 Overall High Level Functionality and Architecture Impacts of Flow Based Charging

TS23.135 Multicall Supplementary Service

TS23.140 Multimedia Messaging Service (MMS); Functional Description

TS23.141 Presence Service; Architecture and Functional Description

TS23.195 Provision of UE Specific Behaviour Information to Network Entities

TS23.205 Bearer Independent Circuit Switched Core Network

TS23.207 End-to-End Quality of Service (QoS) Concept and Architecture

TS23.218 IP Multimedia (IM) Session Handling; IM Call Model

TS23.221 Architectural Requirements

TS23.228 IP Multimedia Subsystem (IMS); Stage 2

TS23.234 3GPP System to Wireless Local Area Network (WLAN) Interworking

TS23.236 Intra Domain Connection of RAN Nodes to Multiple CN Nodes

TS23.240 3GPP Generic User Profile (GUP); Architecture

TS23.246 Multimedia Broadcast/Multicast Service (MBMS); Architecture and Functional Description

TS23.271 Location Services (LCS); Functional Description

TS23.981 Interworking Aspects and Migration Scenarios for IPv4 Based IMS Implementations

TS24.002 GSM – UMTS Public Land Mobile Network (PLMN) Access Reference Configuration

TS24.007 Digital Cellular Telecommunications System (Phase 2+); Mobile Radio Interface Signalling Layer 3 General Aspects

TS24.008 Mobile Radio Interface Layer 3 Specification; Core Network Protocols

TS24.011 Point-to-Point (PP) Short Message Service (SMS) Support on Mobile Radio Interface

TS24.228 Signalling Flow for the IP Multimedia Call Control Based on SIP and SDP

TS24.229 IP Multimedia Call Control Protocol Based on SIP and SDP

TS25.301 Radio Interface Protocol Architecture

TS25.303 UE Functions and Inter-Layer Procedures in Connected Mode

TS25.304 UE Procedures in Idle Mode and Procedures for Cell Reselection in Connected Mode

TS25.331 Radio Resource Control (RRC) Protocol Specification

TS25.401 UTRAN Overall Description

TS25.410 UTRAN Iu Interface: General Aspects and Principles

TS25.413 UTRAN Iu Interface RANAP Signalling

TS25.931 UTRAN Functions, Examples on Signalling Procedures

TS26.141 IP Multimedia System (IMS) Messaging and Presence; Media Formats and Codec's

TS26.235 Packet Switched Multimedia Applications; Default Codecs

TS27.005 Use of Data Terminal Equipment - Data Circuit Terminating Equipment (DTE - DCE) Interface for Short Message Service (SMS) and Cell Broadcast Service (CBS)

TS27.060 Mobile Station (MS) Supporting Packet Switched Services

TS29.002 Mobile Application Part (MAP)

TS29.016 Serving GPRS Support Node (SGSN) – Visitor Location Register (VLR); Gs Interface Network Service Specification

TS29.018 Serving GPRS Support Node (SGSN) – Visitor Location Register (VLR); Gs Interface Layer 3 Specification

TS29.061 Interworking Between the Public Land Mobile Network (PLMN) Supporting Packet-Based Services and Packet Data Networks (PDN)

TS29.207 Policy Control over Go Interface

TS29.208 End-to-End Quality of Service (QoS) Signalling Flows

TS29.228 IP Multimedia (IM) Subsystem Cx and Dx Interfaces; Signalling Flows and Message Contents

TS29.229 Cx and Dx Interfaces Based on the Diameter Protocol, Protocol Details

TS31.101 UICC – Terminal Interface; Physical and Logical Characteristics

TS31.102 Characteristics of the USIM Application

TS32.200 Telecommunication Management; Charging Management; Charging Principles

TS32.225 Telecommunication Management; Charging Management; Charging Data Description for the IP Multimedia Subsystem

TS32.250 Circuit Switched (CS) Domain Charging

TS32.251 Packet Switched (PS) Domain Charging

TS32.252 Wireless Local Area Network (WLAN) Charging

TS32.260 IP Multimedia Subsystem (IMS) Charging

TS32.270 Multimedia Messaging Service (MMS) Charging

TS32.271 Location Services (LCS) Charging

TS32.295 Charging Data Record (CDR) Transfer

TS32.296 Online Charging System (OCS) Applications and Interfaces

TS32.297 Charging Data Record (CDR) File Format and Transfer

TS32.298 Charging Data Record (CDR) Parameter Description

TS32.299 Diameter Charging Application

TS33.102 Security Architecture

TS33.103 Integration Guidelines

TS33.120 Security Principles and Objectives

TS33.141 Presence Service; Security

TS33.203 Access Security for IP-based Services

TS33.210 Network Domain Security; IP Network Layer Security

TS33.222 Generic Authentication Architecture (GAA); Access to Network Application Functions Using Hypertext Transfer Protocol over Transport Layer Security (HTTPS)

TS33.234 WLAN Interworking Security

TS41.001 GSM Release Specifications

TS42.017 Subscriber Identity Modules (SIM); Functional Characteristics

TS43.020 Security Related Network Functions

TS43.047 Example Protocol Stacks for Interconnecting Service Centre(s) (SC) and Mobile-Services Switching Centre(s) (MSC)

TS43.048 Security Mechanisms for the SIM Application Toolkit

TS43.051 Technical Specification Group GSM/EDGE Radio Access Network; Overall Description

TS43.068 Voice Group Call Service (VGCS)

TS43.069 Voice Broadcast Service (VBS)

TS44.006 Mobile Station – Base Station System (MS - BSS) interface; Data Link (DL) Layer Specification

TS44.008 Mobile Radio Interface Layer 3 Specification

TS44.018 Mobile Radio Interface Layer 3 Specification, Radio Resource Control Protocol

TS44.071 Mobile Radio Interface Layer 3 Location Services (LCS) Specification

TS45.008 Radio Subsystem Link Control

TS48.001 Base Station System – Mobile Services Switching Centre (BSS-MSC) Interface; General Aspects

TS48.002 Base Station System – Mobile Services Switching Centre (BSS-MSC) Interface; Interface Principles

TS48.004 Base Station System – Mobile Services Switching Centre (BSS-MSC) Interface Layer 1 Specification

TS48.006 Signalling Transport Mechanism Specification for the Base Station System – Mobile Services Switching Centre (BSS-MSC) Interface

TS48.008 Mobile Services Switching Centre – Base Station System (MSC-BSS) Interface; Layer 3 Specification

TS48.014 Base Station System (BSS) – Serving GPRS Support Node (SGSN) Interface; Gb Interface Layer 1

TS48.016 Base Station System (BSS) – Serving GPRS Support Node (SGSN) Interface; Network Service

TS48.018 Base Station System (BSS) – Serving GPRS Support Node (SGSN) Interface; BSS GPRS Protocol (BSSGP)

TS48.031 Serving Mobile Location Centre – Serving Mobile Location Centre (SMLC-SMLC) ; SMLCPP Specification

TS48.051 Base Station Controller – Base Transceiver Station (BSC-BTS) Interface; General Aspects

TS48.052 Base Station Controller – Base Transceiver Station (BSC-BTS) Interface; Interface Principles

TS48.054 Base Station Controller – Base Transceiver Station (BSC-BTS) Interface; Layer 1 Structure of Physical Circuits

TS48.056 Base Station Controller – Base Transceiver Station (BSC-BTS) Interface; Layer 2 Specification

TS48.058 Base Station Controller – Base Transceiver Station (BSC-BTS) Interface; Layer 3 Specification

TS49.031 Network Location Services (LCS); Base Station Application Part LCS Extension (BSSAP-LE)

TS51.011 Specification of the Subscriber Identity Module - Mobile Equipment (SIM-ME) Interface

IETF IMS-Related Documentation

RFC1889 RTP: A Transport Protocol for Real-time Applications

RFC2246 The TLS Protocol Version 1.0

RFC2373 IP Version 6 Addressing Architecture

RFC2396 Uniform Resource Identifiers (URI): Generic Syntax

RFC2401 Security Architecture for the Internet Protocol

RFC2462 IPv6 Address Autoconfiguration

RFC2486 The Network Access Identifier

RFC2616 Hypertext Transfer Protocol – HTTP/1.1

RFC2617 HTTP Authentication: Basic and Digest Access Authentication

RFC2663 IP Network Address Translator (NAT) Terminology and Considerations

RFC2766 Network Address Translation – Protocol Translation (NAT-PT)

RFC2778 A Model for Presence and Instant Messaging

RFC2779 Instant Messaging/Presence Protocol Requirements

RFC2806 URLs for Telephone Calls

RFC2833 RTP Payload for DTMF Digits, Telephony Tones and Telephony Signals

RFC2893 Transition Mechanisms for IPv6 Hosts and Routers

RFC2916 E.164 Number and DNS

RFC2976 The SIP INFO Method

RFC3041 Privacy Extensions for Stateless Address Autoconfiguration in IPv6

RFC3261 SIP: Session Initiation Protocol

RFC3262 Reliability of Provisional Responses in Session Initiation Protocol (SIP)

RFC3263 Session Initiation Protocol (SIP): Locating SIP Servers

RFC3265 Session Initiation Protocol (SIP) Specific Event Notification

RFC3310 HyperText Transfer Protocol (HTTP) Digest Authentication Using Authentication and Key Agreement (AKA)

RFC3311 The Session Initiation Protocol (SIP) UPDATE Method

RFC3312 Integration of Resource Management and Session Initiation Protocol (SIP)

RFC3313 Private Session Initiation Protocol (SIP) Extensions for Media Authorization

RFC3315 The Session Initiation Protocol (SIP) REFER Method

RFC3316 IPv6 for Some Second and Third Generation Cellular Hosts

RFC3329 (2002) Security Mechanism Agreement for the Session Initiation Protocol (SIP)

RFC3323 (2002) A Privacy Mechanism for the Session Initiation Protocol (SIP)

RFC3325 (2002) Private Extensions to the Session Initiation Protocol (SIP) for Asserted Identity within Trusted Network

RFC3327 Session Initiation Protocol Extension Header for Registering Non-Adjacent Contacts

RFC3329 Security Mechanism Agreement for the Session Initiation Protocol (SIP)

RFC3388 Grouping of Media Lines in Session Description Protocol (SDP)

RFC3428 Session Initiation Protocol (SIP) Extension for Instant Messaging

RFC3455 Private Header (P-Header) Extensions to the Session Initiation Protocol (SIP) for the 3rd Generation Partnership Project (3GPP)

RFC3485 The Session Initiation Protocol (SIP) and Session Description Protocol (SDP) Static Dictionary for Signalling Compression (SIGCOMP)

RFC3486 Compressing the Session Initiation Protocol (SIP)

RFC3515 The Session Initiation Protocol (SIP) REFER Method

RFC3524 Mapping of Media Streams to Resource Reservation Flows

RFC3556 Session Description Protocol (SDP) Bandwidth Modifiers for RTP Control Protocol (RTCP) Bandwidth

RFC3588 Diameter Base Protocol

RFC3608 Session Initiation Protocol (SIP) Extension Header Field for Service Route Discovery During Registration

RFC3680 A Session Initiation Protocol (SIP) Event Package for Registrations

RFC3761 The E.164 to Uniform Resource Identifiers (URI) Dynamic Delegation Discovery System (DDDS) Application (ENUM)

RFC3840 Indicating User Agent Capabilities in the Session Initiation Protocol (SIP)

RFC3841 Caller Preferences for the Session Initiation Protocol (SIP)

RFC3966 The Tel URI for Telephone Numbers

Bibliography

3rd Generation Partnership Project (3GPP), "3G Security; Access Security for IP-based Services," *TS 33.203 V.5.9.0* (2004-09), Release 5.

3rd Generation Partnership Project (3GPP), "3G Security; Security Threats and Requirements," *TS 21.133 V4.1.0* (2001-12), Release 4.

3rd Generation Partnership Project (3GPP), "All-IP Core Network Multimedia Domain, IP Multimedia (IMS) Session Handling; IP Multimedia (IM) Call Model; Stage 2," *X.S0013-003-A, Version 1.0*, November 2005.

3rd Generation Partnership Project (3GPP), "IP Multimedia Subsystem – Accounting Information Flows and Protocols," *X.S0013-008-0, Version 2.0,* July 2005.

3rd Generation Partnership Project (3GPP), "Numbering, Addressing, and Identification," *TS 23.003 v3.10.0* (2002-06), Release 1999.

3rd Generation Partnership Project (3GPP), "Telecommunication Management; Charging Management; Charging Architecture and Principles," *TS 32.240, V6.0.0* (2004-09), Release 6.

3rd Generation Partnership Project (3GPP), "Telecommunication Management; Charging Management; Diameter Charging Applications," *TS 32.299, V6.1.0* (2004-12), Release 6.

3rd Generation Partnership Project (3GPP), "Telecommunication Management; Charging Management; IP Multimedia Subsystem (IMS) Charging," *TS 32.260, V7.0.0* (2006-09), Release 7.

ATIS, "ATIS Next Generation Network (NGN) Framework Part II: NGN Roadmap 2005," Issue 1.0, June 17, 2005.

Calhoun, P., J. Loughney, E. Guttman, G. Zorn, and J. Arkko, "Diameter Base Protocol," *RFC 3588,* Internet Engineering Task Force (IETF), September 2003.

Donovan, S., "The SIP INFO Method," *RFC 2976,* Internet Engineering Task Force (IETF), October 2000.

Droms, R., J. Bound, B. Volz, T. Lemon, Perkins, and M. Carney, "Dynamic Host Configuration Protocol for IPv6 (DHCPv6)," *RFC 3315,* Internet Engineering Task Force (IETF), July 2003.

GSM Association, "IMS Roaming and Interworking Guidelines," *Version 3.4,* January 7, 2006.

GSM Association, "IMS Services and Applications," *Version 1.0,* December 6, 2004.

Hakala, Harri, Leena Mattila, Juha-Pekka Koskinen, Marco Stura, and John Loughney, "Draft-IETF-AAA-Diameter-CC-05," Internet Engineering Task Force (IETF), May 14, 2004.

Halonen, T., J. Romero, and J. Melero, *GSM, GPRS, and EDGE Performance,* John Wiley and Sons, West Sussex, England, 2002.

Hinden, R. and S. Deering, "IP Version 6 Addressing Architecture," *RFC 4291,* Internet Engineering Task Force (IETF), February 2006.

Jennings, C., J. Peterson, and M. Watson, "Private Extensions to the Session Initiation Protocol (SIP) for Asserted Identity within Trusted Networks," *RFC 3325,* Internet Engineering Task Force (IETF), November 2002.

Rosenberg, J., "The Session Initiation Protocol (SIP) Update Method," *RFC 3311,* Internet Engineering Task Force (IETF), September 2002.

Rosenberg, J., H. Schulzrinne, G. Camarillo, A. Johnston, J. Peterson, R. Sparks, M. Handley, and E. Schooler, "SIP: Session Initiation Protocol," *RFC 3261,* Internet Engineering Task Force (IETF), June 2002.

Telcordia Technologies, "Specification of Signalling System Number 7, Incoming Call Interworking from SIP to BICC/ISUP at I-IWU," *GR-246-Core,* Issue 10, December 2005.

Index